John Lubbock

Flowers, Fruits and Leaves

ISBN/EAN: 9783337107536

Printed in Europe, USA, Canada, Australia, Japan

Cover: Foto ©berggeist007 / pixelio.de

More available books at **www.hansebooks.com**

John Lubbock

Flowers, Fruits and Leaves

NATURE SERIES.

FLOWERS, FRUITS,

AND

LEAVES.

BY

SIR JOHN LUBBOCK, Bart., F.R.S., M.P., D.C.L., LL.D.

WITH NUMEROUS ILLUSTRATIONS.

London:

MACMILLAN AND CO.

AND NEW YORK.

1888.

PREFACE.

THE volume of my Scientific Lectures which contains a chapter on Flowers being now almost out of print, Messrs. Macmillan have proposed to me that I should reprint that chapter, together with two subsequent Lectures, one on Fruits and Seeds, the other on Leaves, as an additional volume in their Nature Series, under the title of " Flowers, Fruits, and Leaves." To this I have gladly assented, and have taken the opportunity to make several corrections and additions, as well as to insert several new illustrations.

For permission to use Figs. 80, 82-87, 88-90, and 95, I am indebted to the courtesy of Messrs. Lovell Reeve and Co.

HIGH ELMS, DOWN, KENT.

CONTENTS.

CHAPTER I.

CHAPTER II.

CHAPTER III.

CHAPTER IV.

CHAPTER V.

CHAPTER VI.

LIST OF ILLUSTRATIONS.

FIG. 1.—*Lamium album.*

FLOWERS.

CHAPTER I.

THE flower of the common White Deadnettle (*Lamium album*, Fig. 1) consists of a narrow tube, somewhat expanded at the upper end (Fig. 2), where the lower lobe of the corolla forms a platform, on each side of which is a small projecting lobe (Fig. 3, *m*). The upper portion of the corolla is an arched hood (Fig. 3 *co*), under which lie four anthers (*a a*), in pairs, while between them, and projecting somewhat downwards, is the pointed pistil (*st*). At the lower part, the tube contains honey, and above the honey is a row of hairs almost closing the tube.

Now, why has the flower this peculiar form?

What regulates the length of the tube? What is the use of this arch? What lessons do these lobes teach us? What advantage is the honey to the flower? Of what use is the fringe of hairs? Why does the stigma project beyond the anthers? and why is the corolla white, while the rest of the plant is green?

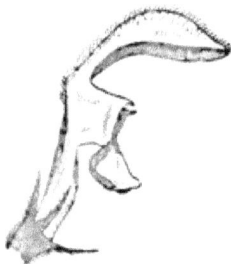

FIG. 2.—Flower of *Lamium album.*   FIG. 3.—Section of ditto.

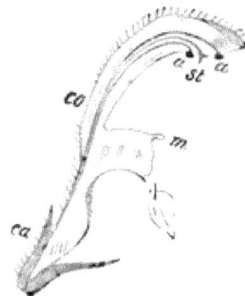

Similar questions may of course be asked with reference to other flowers. Let us now see whether we can throw any light upon them.

At the close of the last century, Conrad Sprengel published a valuable book on flowers, in which he pointed out that the forms and colours, the scent, honey, and general structure of flowers, have reference to the visits of insects, which are of importance in transferring the pollen from the stamens to the pistil. This admirable work, however, did not attract the attention it deserved, and remained almost unknown until Mr. Darwin devoted himself to the subject. Our illustrious countryman was the first clearly to perceive that the essential service which insects perform to flowers consists not only in transferring the pollen

from the stamens to the pistil, but in transferring it
from the stamens of one flower to the pistil of another.
Sprengel had indeed observed in more than one instance that this was the case, but he did not altogether
appreciate the importance of the fact.

Mr. Darwin, however, has not only made it clear
from theoretical considerations, but has also proved it,
in a variety of cases, by actual experiment. More
recently Fritz Müller has even shown that in some
cases pollen, if placed on the stigma of the same
flower, has no more effect than so much inorganic
dust ; while, and this is perhaps even more extra-
ordinary, in others, the pollen placed on the stigma of
the same flower acted on it like a poison. This he
observed in several species ; the flowers faded and fell
off, the pollen masses themselves, and the stigma in
contact with them, shrivelled up, turned brown, and
decayed ; while flowers on the same bunch, which were
left unfertilised, retained their freshness.

The importance of this "cross-fertilisation," as it
may be called, in contradistinction to "self-fertilisation," was first conclusively proved by Mr. Darwin in
his remarkable memoir on Primula (*Linnean Journal*,
1862), and he has since illustrated the same rule by
researches on Orchids, Linum, Lythrum, and a variety
of other plants. The new impulse thus given to the
study of flowers has been followed up in this country
by Hooker, Ogle, Bennett, and other naturalists, and
on the Continent by Axell, Delpino, Hildebrand,
Kerner, F. Müller, and especially by Dr. H. Müller,
who has brought together the observations of others,
and added to them an immense number of his own.

In by far the majority of cases, the relation between flowers and insects is one of mutual advantage. In some plants, however, as for instance in our common Drosera, we find a very different state of things, and the plant catches and devours the insects.[1]   The first observation on insect-eating flowers was made about the year 1768, by our countryman Ellis. He observed that in Dionæa, a North American plant, the leaves

FIG. 4 —*Drosera rotundifolia.*

have a joint in the middle, and thus close over, kill, and actually digest any insect which may alight on them.

In our common Sundew (*Drosera rotundifolia*, Fig. 4) the rounded leaves are covered with glutinous glandular hairs or tentacles—on an average about 200 on a full-sized leaf. The glands are each surrounded

[1] See Darwin's *Insectivorous Plants.*

by a drop of an exceedingly viscid solution, which, glittering in the sun, has given rise to the name of the plant. If any object be placed on the leaf, these glandular hairs slowly fold over it, though if it be inorganic they soon unfold again. On the other hand, if any small insect alights on the leaf it becomes entangled in the glutinous secretion, the glands close over it, their secretion is increased, and they literally digest their prey. Mr. Frank Darwin has recently shown that plants supplied with insects grow more vigorously than those not so fed. It is very curious that while the glands are so sensitive that even an object weighing only $\frac{1}{78740}$ of a grain placed on them is sufficient to cause motion, yet they are "insensible to the weight and repeated blows of drops" of even heavy rain.

Drosera, however, is not our only English insectivorous plant. In the genus Pinguicula, which frequents moist places, the leaves are concave with incurved margins, and the upper surfaces are covered with two sets of glandular hairs. In this case the naturally incurved edges curve over still more if a fly or other insect be placed on the leaf.

Another case is that of Utricularia, an aquatic species, which bears a number of utricles or sacs, which have been supposed to act as floats. Branches, however, which bear no bladder float just as well as the others, and there seems no doubt that their real use is to capture small aquatic animals, which they do in considerable numbers. The bladders in fact act on the principle of an eel-trap, having an orifice closed with a flap which permits an easy entrance, but effectually prevents the unfortunate victim from getting out again.

I will only allude to one foreign case, that of the Sarracenia.[1]   In this genus some of the leaves are in the form of a pitcher.   They secrete a fluid, and are lined internally with hairs pointing downwards. Up the outside of the pitcher there is a line of honey glands which lure the insects to their destruction.   Flies and other insects which fall into this pitcher cannot get out again, and are actually digested by the plant. Bees, however, are said to be scarcely ever caught.

Every one knows how important flowers are to insects ; every one knows that bees, butterflies, etc., derive the main part of their nourishment from the honey or pollen of flowers, but comparatively few are aware, on the other hand, how much the flowers them- selves are dependent on insects.   Yet it has, I think, been clearly shown that if insects have been in some respects modified and adapted with a view to the acquirement of honey and pollen, flowers, on the other hand, owe their scent and honey, their form and colour, to the agency of insects.   Thus the lines and bands by which so many flowers are ornamented have reference to the position of the honey ; and it may be observed that these honey-guides are absent in night flowers, where they of course would not show, and would therefore be useless, as for instance in *Lychnis vespertina* or *Silene nutans.*   Night flowers, moreover, are generally pale ; for instance, *Lychnis vespertina* is white, while *Lychnis diurna*, which flowers by day, is red.

See Hooker, *British Association Journal,* 1874.

Indeed, it may be laid down as a general rule that
those flowers which are not fertilised by insects, as for
instance those of the Beech and most other forest
trees, are small in size, and do not possess either
colour, scent, or honey.

Before proceeding further let me briefly mention
the terms used in describing the different parts of a
flower.

If we examine a common flower, such for instance
as a Geranium, we shall find that it consists, firstly,
of an outer envelope or *calyx*, sometimes tubular,
sometimes consisting of separate leaves called *sepals ;*
secondly, an inner envelope or *corolla*, which is generally more or less coloured, and which, like the calyx,
is sometimes tubular, sometimes composed of separate
leaves called *petals ;* thirdly, of one or more *stamens*,
consisting of a stalk or *filament*, and a head or *anther*,
in which the pollen is produced ; and fourthly, a *pistil*,
which is situated in the centre of the flower, and
consists generally of three principal parts ; one or
more compartments at the base, each containing one
or more seeds; the stalk or style; and the *stigma*,
which in many familiar instances forms a small head
at the top of the style or ovary, and to which the
pollen must find its way in order to fertilise the
flower.

But though the pistil is thus surrounded by one
or more rows of stamens, there are comparatively few
cases in which the pollen of the latter falls directly
on the former. On the contrary this transference is
in most cases effected in other ways—generally by
means of the wind, of insects, or, in some cases, of

birds. In the former case, however, by far the
greater part of the pollen is wasted ; and much more
must therefore be produced than in those cases where
the transference is effected by insects.

One advantage, of course, is the great economy of
pollen. We have not much information on the subject, but it would seem, from the few observations
that have been made, that half a dozen pollen grains
are sufficient to fertilise a seed. But in plants in

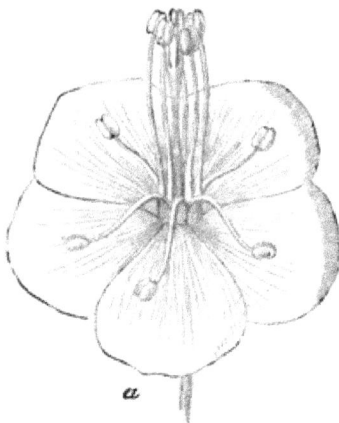

FIG. 5.—*Geranium pratense* (young
flower). Five of the stamens are
erect.

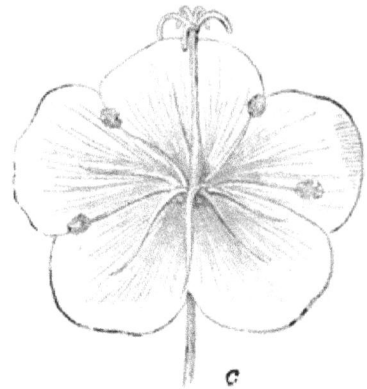

FIG. 6.—*Geranium pratense* (older
flower). The stamens have retired,
and the stigmas are expanded.

which the pollen is carried by the wind, the chances
against any given grain reaching the pistil of another
flower are immense. Consequently by far the greater
part of the pollen is lost. Every one, for instance,
must have observed the clouds of pollen produced by
the Scotch fir. In such flowers as the Pæony the
pollen is carried by insects, and far less therefore is
required ; yet even here the quantity produced is still
large ; it has been estimated that each flower produces

between 3,000,000 and 4,000,000 grains of pollen. The Dandelion is more specialised in this respect, and produces far less pollen; according to Mr. Hassall about 240,000 grains to each flower; while in *Geum urbanum*, according to Gærtner, only ten times more pollen is produced than is actually used in fertilisation.

It might, however, be at first supposed that where stamens and pistil co-exist in the same flower, the pollen from the one could easily fall on and fertilise the other. And in fact in some species this does occur; but as we have seen, it is a great advantage to a species that the flower should be fertilised by pollen from a different stock. How then is self-fertilisation prevented?

There are three principal modes.

Firstly, in many species the stamens and pistil are in separate flowers, sometimes situated on different plants.

Secondly, even when the stamens and pistil are in the same flower, they are in many species not mature at the same time; this was first observed by Sprengel in *Epilobium angustifolium* as long ago as 1790; in some cases the stigma has matured before the anthers are ripe, while in other and more numerous cases the anthers have ripened and shed all their pollen before the stigma has come to maturity.

Thirdly, there are many species in which, though the anthers and stigma are contained in the same flower and are mature at the same time, they are so situated that the pollen can hardly reach the stigma of the same flower.

The transference of the pollen from one flower to another is, as already mentioned, effected principally either by the wind or by insects, though in some cases it is effected by other agencies, as for instance, by birds, or by water. For instance, in the curious *Vallisneria spiralis* (Fig. 36) the female flowers are situated on long stalks which are spirally twisted, and grow very rapidly, so that even if the level of the water alters, provided this be within certain limits, the flowers float on the surface. The male flowers on the contrary are minute and sessile, but when mature they detach themselves from the plant, rise to the surface, and float about freely like little boats among the female flowers.

Wind-fertilised flowers as a rule have no colour, emit no scent, produce no honey, and are symmetrical in form. Colour, scent, and honey are the three characteristics by which insects are attracted to flowers.

As already observed, wind-fertilised flowers generally produce much more pollen than those which are fertilised by insects. This is necessary, because it is obvious that the chances against any given pollen grain reaching the stigma are much greater in the one case than in the other.

Again, it is an advantage to wind-fertilised plants to flower early in the spring before the leaves are out, because the latter would catch much of the pollen and thus interfere with its access to the stigma. In these plants the pollen is less adherent, so that it can easily be blown away by the wind, which would be a disadvantage in most plants which are

fertilised by insects. Such flowers generally have the
stigma more or less branched or hairy, which evidently
must tend to increase their chances of catching the
pollen.

Moreover, as Mr. Darwin has observed, there does
not appear to be a single instance of an irregular
flower which is not fertilised by insects or birds.

The evidence derivable from the relations of bees
to flowers is probably sufficient to satisfy most
minds that bees are capable of distinguishing colours,
but the fact had not been proved by any conclusive
experiments. I therefore tried the following. If you
bring a bee to some honey, she feeds quietly, goes
back to the hive, stores away her honey, and returns
with or without companions for another supply.
Each visit occupies about six minutes, so that there
are about ten in an hour, and about a hundred in a
day. I may add that in this respect the habits of
wasps are very similar, and that they appear to be
quite as industrious as bees.

I once tested this by training a bee and a wasp to
come to some honey, and then timing them through
a whole day. Knowing they would be early I went
into my study a few minutes after four in the morning,
but the wasp was already at work, and continued
without a moment's intermission until 7.46 in the
evening, working without a moment's rest for nearly
sixteen hours and making no less than 116 visits to
the honey. The bee began at 5.45 A.M., or somewhat
later than the wasp, and left off also rather earlier.
Perhaps I may give the record of the morning's work
of this wasp.

At each visit she sucked up as much honey as she could carry off to the nest and returned at once. She came, as already mentioned, for the first time,

At 4.13, returning at		At 8.29, returning at	
4.32	,,	8.36	,,
4.50	,,	8.40	,,
5.5	,,	8.45	,,
5.15	,,	8.56	,,
5.22	,,	9.7	,,
5.29	,,	9.14	,,
5.36	,,	9.20	,,
5.43	,,	9.26	,,
5.50	,,	9.39	,,
5.57	,,	9.43	,,
6.5	,,	9.50	,,
6.14	,,	9.57	,,
6.23	,,	10.4	,,
6.30	,,	10.10	,,
6.40	,,	10.15	,,
6.48	,,	10.24	,,
6.56	,,	10.29	,,
7.5	,,	10.37	,,
7.12	,,	10.45	,,
7.18	,,	10.50	,,
7.25	,,	10.59	,,
7.31	,,	11.6	,,
7.40	,,	11.15	,,
7.46	,,	11.22	,,
7.52	,,	11.30	,,
8.0	,,	11.35	,,
8.10	,,	11.47	,,
8.18	,,	11.55	,,
8.24	,,	12.6	,,

The visits were continued till dusk.

This, however, was in autumn ; in summer they make more overtime, and work on even late in the evening.

In fine weather bees visit often more than twenty
flowers in a minute, and so carefully do they economise
the sunny hours, that in flowers with several nectaries
if they find one dry, they do not waste time by ex-
amining the others on the same plant. Mr. Darwin
watched carefully certain flowers, and satisfied himself that each one was visited by bees at least thirty
times in a day. The result is, that even where flowers
are very numerous—as, for instance, on heathy plains
and clover fields—every one is visited during the day.
Mr. Darwin has carefully examined a large number of
flowers in such cases, and found that every single one
had been visited by bees.

In order to test the power of bees to appreciate
colour, I placed some honey on a slip of glass, and
put the glass on coloured paper. For instance, I put
some honey in this manner on a piece of blue paper,
and when a bee had made several journeys, and thus
become accustomed to the blue colour, I placed some
more honey in the same manner on orange paper
about a foot away. Then during one of the absences
of the bee I transposed the two colours, leaving the
honey itself in the same place as before. The bee
returned as usual to the place where she had been
accustomed to find the honey; but though it was
still there, she did not alight, but paused for a
moment, and then dashed straight away to the blue
paper. No one who saw my bee at that moment
could have had the slightest doubt of her power of
distinguishing blue from orange.

Again, having accustomed a bee to come to honey
on blue paper, I ranged in a row other supplies of

honey on glass slips placed over paper of other colours, yellow, orange, red, green, black, and white. Then I continually transposed the coloured paper, leaving the honey on the same spots; but the bee always flew to the blue paper, wherever it might be. Bees appear fortunately to prefer the same colours as we do. On the contrary, flowers of a livid yellow, or fleshy colour are most attractive to flies; and moreover while bees are attracted by odours which are also agreeable to us, flies, as might naturally be

Fig. 7.—*Malva sylvestris.*

Fig. 8.—*Malva rotundifolia.*

expected from the habits of their larvæ, prefer some which to us seem anything but pleasant.

Among other obvious evidences that the beauty of flowers is useful in consequence of its attracting insects, we may adduce those cases in which the transference of the pollen is effected in different manners in nearly allied plants, sometimes even in the same genus.

Thus, as Dr. H. Müller has pointed out, *Malva*

sylvestris (Fig. 7) and *Malva rotundifolia* (Fig. 8)
which grow in the same localities, and therefore must
come into competition, are nevertheless nearly equally
common.

In *Malva sylvestris*, however (Fig. 9), where the
branches of the stigma are so arranged that the plant
cannot fertilise itself, the petals are large and conspicuous, so that the plant is visited by numerous insects ;
while in *Malva rotundifolia*, the flowers of which are
comparatively small and rarely visited by insects, the

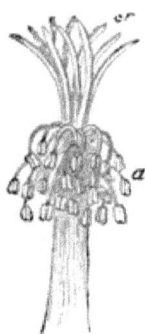

FIG 9 —Stamens and stigmas of   FIG. 10.—Ditto of *Malva rotundifolia*
    Malva sylvestris.

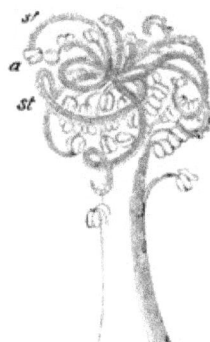

branches of the stigma are elongated, and twine themselves (Fig. 10) among the stamens, so that the flower
readily fertilises itself.

Another interesting case is afforded by the genus
Epilobium. *Epilobium angustifolium* has large pur-
plish flowers in conspicuous heads (Fig. 11), and is
much frequented by insects; while *E. parviflorum*
(Fig. 12) has small solitary flowers and is seldom
visited by insects. Now in the former species their
visits are necessary, because the stamens ripen and

shed their pollen before the pistil, so that the flower is consequently incapable of fertilising itself. In *E. parviflorum*, on the contrary, the stamens and pistil come to maturity at the same time.

Let us take another case—that of certain Geraniums. In *G. pratense* (Figs. 5 and 6, p. 8) all the stamens open, shed their pollen, and wither away before the pistil comes to maturity. The flower cannot therefore fertilise itself, and depends entirely on the visits of

Fig. 11.—*Epilobium angustifolium.*   Fig. 12.—*Epilobium parviflorum.*

insects for the transference of the pollen. In *G. pyrenaicum*, where the flower is not quite so large, all the stamens ripen before the stigma, but the interval is shorter, and the stigma is mature before all the anthers have shed their pollen. It is therefore not absolutely dependent on insects. In *G. molle*, which has a still smaller flower, five of the stamens come to maturity before the stigma, but the last five ripen simultaneously with it. Lastly, in *G. pusillum*, which is least of all, the stigma ripens even before the stamens. Thus, then, we have a series more or less dependent on insects, from *G. pratense* to which they are necessary, to *G. pusillum*, which is quite independent of them ; while the size of the corolla increases with the dependence on insects.

In those species in which self-fertilisation is prevented by the circumstance that the stamens and pistil do not come to maturity at the same time, the stamens generally ripen first.

The advantage of this is probably connected with the visits of bees. In those flowers which grow in bunches the lower ones naturally open first. Consequently in any given spike the flowers are at first all male ; subsequently the lower ones, being the older, have arrived at the female stage, while the upper ones are still male. Now it is the habit of bees to begin with the lower flowers of a spike and work upwards. A bee, therefore, which has already dusted herself with pollen from another flower, first comes in contact with the female flowers, and dusts them with pollen, after which she receives a fresh supply from the upper male flowers, with which she flies to another plant.

There are, however, some few species in which the pistil ripens before the stamens. One is our common *Scrophularia nodosa*. Now why is this? Mr. Wilson has given us the answer. *S. nodosa* is one of our few flowers specially visited by wasps ; the honey being not pleasing to bees. Wasps, however, unlike bees, generally begin with the upper flowers and pass downwards, and consequently in wasp flowers it is an advantage that the pistil should ripen before the stamens. But though the stamens generally ripen before the pistil, the reverse sometimes occurs. Of this a very interesting case is that of the genus Aristolochia. The flower is a long tube, with a narrow opening closed by stiff hairs which point backwards, so that it much re- sembles an ordinary eel-trap. Small flies enter the tube in search of honey, but from the direction of the hairs it is impossible for them to return. Thus they are imprisoned in the flower, until the stamens have ripened and shed their pollen, by which the flies get

Fig. 13.—Diagrammatic section of Arum. *h*, hairs ; *a*, anthers ; *st*, stigmas.

thoroughly dusted. Then the hairs of the tube shrivel up, thus releasing the prisoners, which carry the pollen to another flower.

Again, in our common Arums—the Lords and Ladies of village children—the well-known green leaf incloses a central pillar, near the base of which are arranged a number of stigmas (*st*, Fig. 13), and above

them several rows of anthers (*a*). It might be supposed therefore that the pollen from the anthers
would fall on and fertilise the stigmas. This, however, is not what occurs. In fact the stigmas come
to maturity first, and have lost the possibility of fertilisation before the pollen is ripe. The pollen must
therefore be brought by insects, and this is effected by
small flies, which enter the leaf, either for the sake of
honey or of shelter, and which, moreover, when they
have once entered the tube, are imprisoned by the
fringe of hairs (*h*). When the anthers ripen, the pollen falls on to the flies, which in their efforts to escape
get thoroughly dusted with it. Then the fringe of
hairs withers, and the flies, thus set free, soon come
out, and ere long carry the pollen to another plant.

Now let us return to our White Deadnettle and see
how far we can answer the questions which I began
by asking.

In the first place, the honey attracts insects. If
there were no honey, they would have no object in
visiting the flower. The bright colour is useful in
rendering the flower conspicuous. The platform serves
as an alighting stage for bees. The length of the tube
has reference to that of their proboscis, and prevents
the smaller species from obtaining access to the honey
which would be injurious to the flower, as it would
remove the source of attraction for the bees, without
effecting the object in view. The upper arch of the
flower protects the stamens and pistil, and also presses
them firmly against the back of the bee; so that,
when the bee alights on the stage and pushes its proboscis down to the honey, its back comes into contact

with them. The row of small hairs at the bottom of the tube prevents small insects from creeping down the tube and stealing the honey. Lastly, the small processes on each side of the lower lip are the rudimentary representatives of parts, formerly more largely developed, but which, having become useless, have almost disappeared.

In the Deadnettle, it would appear that the pistil matures as early as the stamens, and that cross-fertilisation is attained by the relative position of the stigma, which, as will be seen in the figure, hangs down below the stamens ; so that a bee, bearing pollen on its back from a previous visit to another flower, would touch the pistil and transfer to it some of the pollen, before coming in contact with the stamens. In other species belonging to the same great group (Labiatæ) as Lamium, the same object is secured by the fact that the stamens come to maturity before the pistil ; they shed their pollen, and shrivel up before the stigma is mature.

Fig. 14 represents a young flower of *Salvia officinalis* in which the stamens (*a a*) are mature, but not the pistil (*p*), which, moreover, from its position, is untouched by bees visiting the flower ; as shown in Fig. 15. The anthers, as they shed their pollen, gradually shrivel up ; while, on the other hand, the pistil increases in length and curves downwards, until it assumes the position shown in Fig. 16, *st*, where, as is evident, it must come in contact with any bee visiting the flower, and would touch just that part of the back on which pollen would be deposited by a younger flower. In this manner cross-fertilisation is effectually secured.

There are, however, several other curious points in which *S. officinalis* differs greatly from the species last described.

Fig. 14.

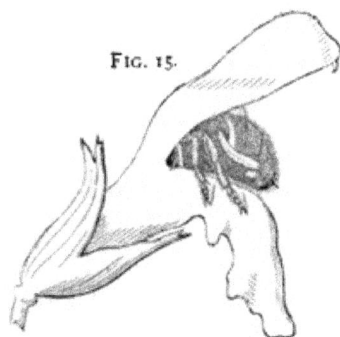

Fig. 15.

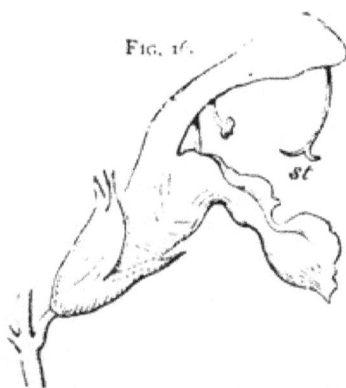

Fig. 16.

Fig. 14.—*Salvia officinalis.* Section of a young flower.
Fig. 15.—Ditto, visited by a Bee.
Fig. 16.—Ditto, older flower.

The general form of the flower, indeed, is very similar. We find again that, as generally in the Labiates, the corolla has the lower lip adapted as an alighting board for insects, while the arched upper lip covers and protects the stamens and pistils.

The arrangement and structure of the stamens is, however, very peculiar and interesting. As in Lamium, they are four in number, but one pair is quite rudimentary (Fig. 14). In the other (*a a*) the two anthers, instead of being attached close together at the summit of the filament, are separated by a long movable rod, or connective (Figs. 17, 18, *m*), so that they can play freely on the stalk of the stamen. In a natural position, this connective is upright, so that the one

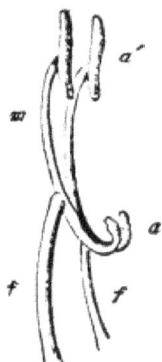

FIG. 17.—Stamens in their natural position.

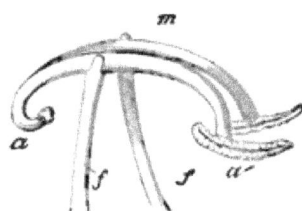

FIG. 18.—Stamens when moved by a Bee.

anther is situated (Fig. 14) in the neck of the tube, the other under the arched hood. The lower anther, moreover, is more or less rudimentary. Now when a bee comes to suck the honey, it pushes the lower anther out of the way with its head ; the result of which is that the connective swings round, and the upper fertile anther comes down on to the back of the bee (Figs. 15 and 18), and dusts it with pollen, just at the place where, in an older flower (Fig. 16) it would be touched by the stigma, *st.*

Fig. 13.—Wild Chervil (*Chærophyllum sylvestre*)

CHAPTER II.

At first sight, it may seem an objection to the view set forward in the preceding chapter, that some flowers—as, for instance, those of the common Antirrhinum—which, according to the above-given tests, ought to be fertilised by insects, are entirely closed. A little consideration, however, will suggest the reply. The Antirrhinum is especially adapted for fertilisation by humble bees. The stamens and pistil are so arranged that smaller species would not effect the object. It is therefore an advantage that they should be excluded, and in fact they are not strong enough to move the spring. The Antirrhinum is, so to say, a closed box, of which the humble bees alone possess the key.

The common Heath (*Erica tetralix*) offers us a very ingenious arrangement. The flower is in the form of

an inverted bell.   The pistil represents the clapper,
and projects a little beyond the mouth of the bell.
The stamens are eight in number, and form a circle
round it, the anthers being united by their sides into
a continuous ring.   Each anther has a lateral hole,
but as long as they touch one another, the pollen
cannot drop out.   Each also sends out a long process,
so that the ring of anthers is surrounded by a row of
spokes.   Now when a bee comes to suck the honey,
it first touches the end of the pistil, on which it could
hardly fail to deposit some pollen had it previously
visited another plant.   It then presses its proboscis
up the bell, in doing which it would pass between
two of the spokes, and pressing them apart, would
dislocate the ring of anthers : a shower of pollen
would thus fall from the open cells on to the head of
the bee.

   In many cases the effect of the colouring and scent
is greatly enhanced by the association of several
flowers in one bunch, or raceme ; as for instance in
the wild hyacinth, the lilac, and other familiar species.
In the great family of Umbelliferæ, this arrangement
is still further taken advantage of, as in the common
Wild Chervil (*Chærophyllum sylvestre*, Fig. 19).

   In this group the honey is not, as in the flowers
just described, situated at the bottom of a tube, but
lies exposed, and is therefore accessible to a great
variety of small insects.   The union of the florets
into a head, moreover, not only renders them more
conspicuous, but also enables the insects to visit a
greater number of flowers in a given time.

   It might at first be supposed that in such small

flowers as these self-fertilisation would be almost un-
avoidable. In most cases, however, the stamens ripen
before the stigmas.

The position of the honey on the surface of a more
or less flat disc renders it much more accessible than
in those cases in which it is situated at the end of a
more or less long tube. That of the Deadnettle, for
instance, is only accessible to certain humble-bees;
while H. Müller has recorded no less than seventy-
three species of insects as visiting the common
Chervil, and some plants are frequented by even a
larger number.

In the Composites, to which the common Daisy
and the Dandelion belong, the association of flowers
is carried so far, that a whole group of florets is ordi-
narily spoken of as one flower. Let us take, for
instance, the common Feverfew, or large white Daisy
(*Chrysanthemum parthenium*, Figs. 20—22). Each
head consists of an outer row of female florets, in
which the tubular corolla terminates on its outer side
in a white leaf-ovary, which serves to make the flower
more conspicuous, and thus to attract insects. The
central florets are tubular, and make up the central
yellow part of the flower-head. Each of these florets
contains a circle of stamens, the upper portions of
which are united at their edges and at the top
(Fig. 0), so as to form a tube, within which is the
pistil. The anthers open inwards, so as to shed the
pollen into this box, the lower part of which is formed
by the stigma, or upper part of the pistil. As the
latter elongates, it presses the pollen against the upper
part of the box, which at length is forced open, and

the pollen is pushed out (Fig. 21). Any insect then
alighting on the flower would carry off some of the
pollen adhering to its under side. The upper part of
the pistil terminates in two branches (Fig. 22, *st*),
each of which bears a little brush of hairs. These
hairs serve to brush the pollen out of the tube ; while

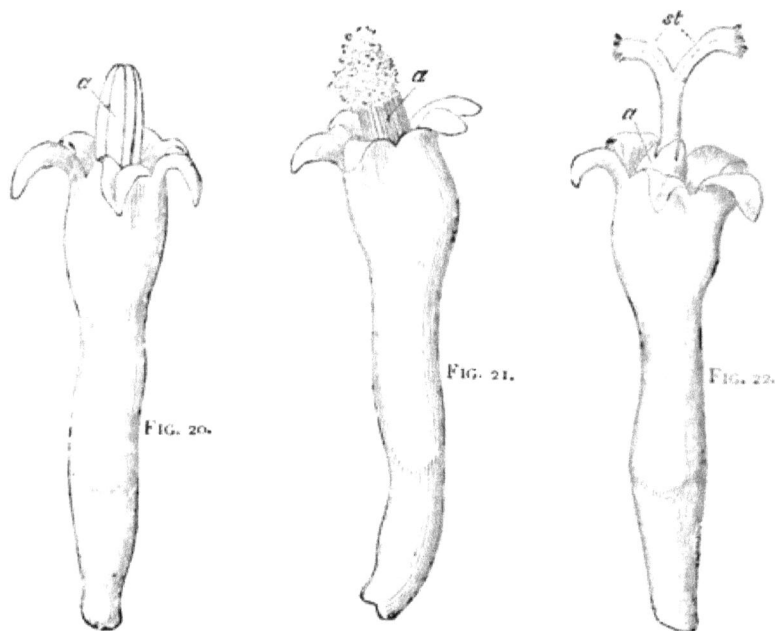

FIG. 20.—Floret of *Chrysanthemum parthenium*, just opened, × 20.
FIG. 21.—Ditto, somewhat more advanced.
FIG. 22.—Ditto, with the stigmas expanded.

in the tube the two branches are pressed close together, but at a later stage they separate, and thus
expose the stigmatic surfaces (Fig. 21), on which an
insect, coming from a younger flower, could hardly
fail to deposit some pollen. The two stigmas in the
ray florets of Parthenium have no brush of hairs ;

and they would be of no use, as these flowers have no stamens.

The Leguminosæ, or Pea-tribe, present a number of beautiful contrivances. Let us take a common little *Lotus corniculatus* (Fig. 23). The petals are five in number; the upper one stands upright, and is known as the standard (Fig. 24, *std*); the two lateral ones present a slight resemblance to wings (Figs. 24, 25, *w*)

Fig. 23.—*Lotus corniculatus.*

while the two lower ones are united along their edges, so as to form a sort of boat, whence they are known as the "keel" (Figs. 25, 26 *k*). The stamens, with one exception, are united at their bases, thus forming a tube (Figs. 27, 28 *t*), surrounding the pistil, which projects beyond them into a triangular space at the end of the keel. Into this space the pollen is shed

(Fig. 28, p°). It must also be observed that each of the wings has a projection (*c*) which locks into a corresponding depression of the keel, so that if the wings are depressed they carry the keel with them. Now when an insect alights on the flower, its weight depresses the wings, and as they again carry with them the keel, the latter slips over the column of stamens thus forcing some of the pollen out at the end of the keel and against the breast of the insect. As soon as the insect leaves the flower, this resumes its natural position, and the pollen is again snugly protected. The arrangement in the Sweet Pea is very similar, and if the wings are seized by the fingers, and pressed down, this out-pumping of the pollen may be easily effected, and the mechanism will then be more clearly understood.

It will be observed (Fig. 28) that one stamen is separated from the rest. The advantage of this is that it leaves a space through which the proboscis of the bee can reach the honey, which is situated inside the tube formed by the united stamens. In those Leguminosæ which have no honey, the stamens are all united together. Such flowers are, nevertheless, in spite of the absence of honey, visited by insects for the sake of the pollen.

In other Leguminosæ, as, for instance, in the Furze (*Ulex europæus*), and the Broom (*Sarothamnus scoparius*), the flower is in a state of tension, but the different parts are, as it were, locked together. The action of the bee, however, puts an end to this; the flower explodes, and thus dusts the bee with pollen.

Whole volumes might be filled with the various

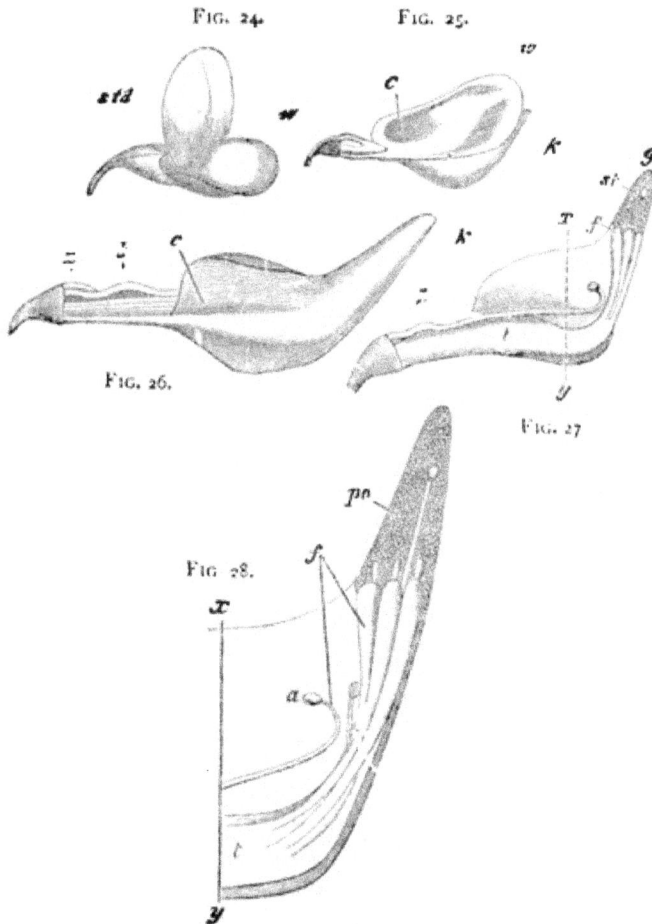

FIG. 24.—Flower of *Lotus corniculatus* seen from the side and in front.
FIG. 25.—Ditto, after removal of the standard.
FIG. 26.—Ditto, after removal of the standard and wings.
FIG. 27.—Ditto, after removal of one side of the keel.
FIG. 28.—Terminal portion of Fig. 27 more magnified.
    a, the free stamen; *c*, the place where the wings lock with the keel;
      f, filaments of stamens; *g*, tip of keel; *po*, pollen; *st*, stigma;
      std, standard; *w*, wing; *k*, keel; *t*, combined bases of stamens.

interesting arrangements by which cross-fertilisation
is secured in this great order of plants.

It is indeed impossible not to be struck by the

marvellous variety of contrivances found among
flowers, and the light thus thrown upon them, by
the consideration of their relations to insects ; but I
must now call attention to certain very curious cases,
in which the same species has two or more kinds of
flowers. Probably in all plants the flowers differ
somewhat in size, and in some species there are two
distinct classes of flowers, one large, and much visited
by insects, the other small, and comparatively neg-
lected ; while in others, as, for instance, some of the
Violets, these differences are carried much further. The
smaller flowers have no scent or honey, the corolla is
rudimentary, and, in fact, an ordinary observer would
not recognise them as flowers at all. Such "cleisto-
gamic" flowers, as they have been termed by Dr.
Kuhn, are already known to exist in over fifty genera.
Their object probably is to secure, with as little expenditure as possible, the continuance of the species
in cases when, from unfavourable weather or other
causes, insects are absent ; and under such circumstances, as scent, honey, and colour are of no use, it is
an advantage to the plant to be spared from the effort
of their production.

As the type of another class of cases in which two
kinds of flowers are produced by the same species
(though not on the same stock) we may take our
common Cowslips and Primroses. If you examine a
number of them, you will find that they fall into two
distinct series. In some of the flowers, the pistil is as
long as the tube, and the button-shaped stigma (Fig.
29, *st*) is situated at the mouth of the flower ; the
stamens (*a a*) being half-way down the tube ; while

in the other set, on the contrary, the anthers are at the
mouth of the flower, and the stigma half-way down.
The existence of these two kinds of flowers had long
been known, but it remained unexplained until Mr.
Darwin devoted his attention to the subject. Now
that he has furnished us with the clue the case is
clear enough.

An insect visiting a plant of the short-styled form
would dust its proboscis at a certain distance from the
extremity (Fig. 30, *a*), which, when the insect passed

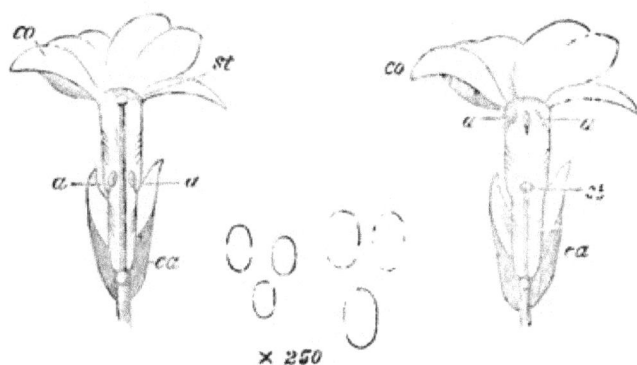

× 250

Fig. 29.—Primula (long-styled form). Fig. 30.—Primula (short-styled form).

to a long-styled flower, would come just opposite to
the pistil (Fig. 29, *st*). At the same time, the stamens
of this second form (Fig. 29, *a*) would dust the proboscis at a point considerably nearer to the extremity,
which in its turn would correspond to the position of
the stigma in the first form (Fig. 30, *st*). The two
kinds of flowers never grow together on the same stock,
and the two kinds of plants generally grow together
in nearly equal proportions. Owing to this arrangement, therefore, insects can hardly fail to fertilise each
flower with pollen from a different stock.

The two forms differ also in some other respects. In the long-styled form, the stigma (*st*) is globular and rough, while that of the short-styled is smoother, and somewhat depressed. These differences, however, are not sufficiently conspicuous to be shown in the figure. Again, the pollen of the long-styled form is considerably smaller than the other, a difference, the importance of which is obvious, for each pollen grain has to give rise to a tube which penetrates the whole length of the style, from the stigma to the base of the flower; and the one has therefore to produce a tube nearly twice as long as that of the other. The careful experiments made by Mr. Darwin have shown that, to obtain the largest quantity of seed, the flowers must be fertilised by pollen from the other form. Nay, in some cases, the flowers produce more seed, if fertilised by pollen from another species, than by that from the other form of their own.

This curious difference in the Primrose and Cowslip, between flowers of the same species, which Mr. Darwin has proposed to call Dimorphism, is found in most species of the genus Primula, but not in all.

The Cowslip and Primrose resemble one another in many respects, but the honey they secrete must be very different, for while the Cowslip is habitually visited during the day by humble-bees, this is not the case with the Primrose, which, in Mr. Darwin's opinion, is fertilised almost exclusively by moths.

The genus Lythrum affords a still more complex case, for here we have three sets of flowers. The stamens are in two groups; in some plants, the pistil projects beyond them; in the second form it is shorter

than any of the stamens, and in the third it is inter-
mediate in length, so that the stigma lies between the
two sets of anthers.

The real use of honey now seems so obvious that it
is curious to see the various theories which were once
entertained on the subject. Patrick Blair thought that
the honey absorbed the pollen, and then fertilised the
ovary. Pontedera thought it kept the ovary in a
moist condition. Linnæus confessed his inability to
solve the question. Other botanists considered that
it was useless material thrown off in the process of
growth. Krünitz thought he observed that in meadows
much visited by bees the plants were more healthy,
but the inference he drew was, that the honey, unless
removed, was very injurious, and that the bees were of
use in carrying it off.

Kurr observed that the formation of honey in
flowers is intimately associated with the maturity of
the stamens and pistil. He lays it down, as a general
rule, that it very seldom commences before the opening of the anthers, is generally most copious during
their maturity, and ceases so soon as the stamens
begin to wither and the development of the fruit commences. Rothe's observations also led him to a
similar conclusion, and yet neither of these botanists
perceived the intimate association which exists
between the presence of honey and the period at
which the visits of insects are of importance to the
plant. Sprengel was the first to point out the real
office of honey, but his views were far from meeting
with general assent, and, even as lately as 1833,
were altogether rejected by Kurr, who came to the

D

conclusion that the secretion of honey is the result of developmental energy, which afterwards concentrates itself on the ovary.

No doubt, however, seems any longer to exist that Sprengel's view is right ; and that the true function of honey is to attract insects, and thus to secure cross-fertilisation. Thus, most of the Rosaceæ are fertilised by insects, and possess nectaries ; but, as Delpino has pointed out, the genus Poterium is anemophilous, or wind-fertilised, and possesses no honey. So also the Maples are almost all fertilised by insects, and produce honey ; but *Acer negundo* is anemophilous, and honey-less. Again, among the Polygonaceæ, some species are insect-fertilised and melliferous, while, on the other hand, certain genera, Rumex and Oxyria, have no honey, and are fertilised by the wind. At first sight it might appear an objection to this view, and one reason perhaps why the earlier botanists missed the true explanation, may have been the fact, that some plants secrete honey on other parts than the flowers. Belt and Delpino have, I think, suggested the true function of these extra floral nectaries.[1] The former of these excellent observers describes a South American species of Acacia : this tree, if unprotected, is apt to be stripped of the leaves by a leaf-cutting ant, which uses them, not directly for food, but, according to Mr. Belt, to grow mushrooms on. The Acacia, however, bears hollow thorns, while each leaf-let produces honey in a crater-formed gland at the

[1] I, by no means, however, wish to suggest that we as yet fully understand the facts. For instance, the use of the nectary at the base of the leaf of the Fern is still quite unexplained.

base, and a small, sweet, pear-shaped body at the tip. In consequence, it is inhabited by myriads of a small ant, which nests in the hollow thorns, and thus finds meat, drink, and lodging all provided for it.   These ants are continually roaming over the plant, and constitute a most efficient body-guard, not only driving off the leaf-cutting ants, but, in Belt's opinion, rendering the leaves less liable to be eaten by herbivorous mammalia.   Delpino mentions that on one occasion he was gathering a flower of *Clerodendron fragrans*, when he was suddenly attacked by a whole army of small ants.

I am not aware that any of our English plants are protected in this manner from browsing quadrupeds, but not the less do our ants perform for them a very similar function, by keeping down the number of small insects, which would otherwise rob them of their sap and strip them of their leaves.

Forel watched, from this point of view, a nest of *Formica pratensis*. He found that the ants brought in dead insects, small caterpillars, grasshoppers, cercopis, &c., at the rate of about twenty-eight a minute, or more than one thousand six hundred in an hour. When it is considered that the ants work not only all day, but in warm weather often all night too, it is easy to see how important a function they fulfil in keeping down the number of small insects.

Some of the most mischievous insects, indeed—certain species, for instance, of aphis and coccus—have turned the tables on the plants, and converted ants from enemies into friends, by themselves developing nectaries, and secreting honey, which the ants

love. We have all seen the little brown garden ant,
for instance, assiduously running up the stems of
plants, to milk their curious little cattle. In this
manner, not only do the aphides and cocci secure
immunity from the attacks of the ants, but even turn
them from foes into friends. They are subject to the
attacks of a species of ichneumon, which lays its eggs
in them, and Delpino has seen ants watching over the
cocci with truly maternal vigilance, and driving off the
ichneumons whenever they attempted to approach.

But though ants are in some respects very useful to
plants, they are not wanted in the flowers. The great
object is to secure cross-fertilisation ; but for this
purpose winged insects are almost necessary, because
they fly readily from one plant to another, and
generally, as already mentioned, confine themselves
for a certain time to the same species. Creeping
insects, on the other hand, naturally would pass from
each floret to the next ; and, as Mr. Darwin has shown
in his last work, it is of little use to bring pollen from
a different flower of the same stock ; it must be from
a different plant altogether. Moreover, creeping insects, in quitting a plant, would generally go up another
close by, without any regard to species. Hence, even
to small flowers (such as many cruciferæ, compositæ,
saxifrages, &c.), which, as far as size is concerned,
might well be fertilised by ants, the visits of flying
insects are much more advantageous. Moreover, if
larger flowers were visited by ants, not only would
these deprive the flowers of their honey, without ful-
filling any useful function in return, but they would
probably prevent the really useful visits of bees. If

you touch an ant with a needle or a bristle, she is almost sure to seize it in her jaws ; and if bees, when visiting any particular species, were liable to have the delicate tip of their proboscis seized on by the horny jaws of an ant, we may be sure that such a plant would soon be deserted.

On the other hand, we know how fond ants are of honey, and how zealously and unremittingly they search for food. How is it, then, that they do not anticipate the bees, and secure the honey for themselves ? Kerner has recently published a most interesting memoir on this subject, and has pointed out a number of ingenious contrivances by which flowers protect themselves from the unwelcome visits of such intruders. The most frequent are the interposition of *chevaux de frise*, which ants cannot penetrate, glutinous surfaces which they cannot traverse, slippery slopes which they cannot climb, or barriers which close the way.

Firstly, then, as regards *chevaux de frise*. In some respects these are the most effectual protection, since they exclude not only creeping insects, but also other creatures, such as slugs. With this object, it will be observed that the hairs which cover the stalks of so many herbs usually point downwards. A good example of this is afforded, for instance, by a plant, *Knautia dipsacifolia* (Fig. 31), allied to our common blue scabious. The heads of the common carline, *Carlina vulgaris* (Fig. 33), again, present a sort of thicket, which must offer an almost impenetrable barrier to ants. Some species of plants are quite smooth, excepting just below the flowers. The common but beautiful Cornflower (*Centaurea cyanus*) is quite smooth, but the involucres forming the flower-head are bordered with recurved teeth. In this case, neither the stem nor the leaves show a trace of such prickles.

The same consideration throws light on the large number of plants which are more or less glutinous, a condition generally produced, as, for instance. in the

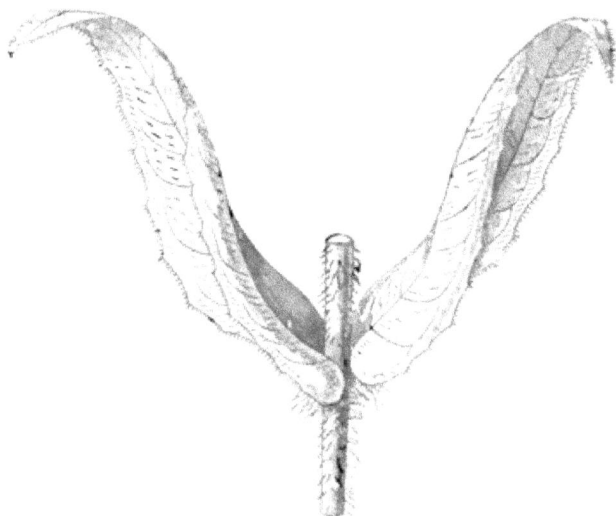

Fig. 31.—*Knautia dipsacifolia.*

flowers of the Gooseberry and of *Linnæa borealis* (Fig. 32), by the presence of glandular hairs. Kerner has called attention to a very interesting illustration afforded by *Polygonum amphibium*. In this species the stigma projects about one-fifth of an inch above the flower, so that if ants could obtain access, they would steal the honey without fertilising the flower; a flying insect, on the contrary, alighting on the flower, could scarcely fail to touch the stigma.

The beautiful rosy flowers of this species are rich in nectar : the stamens are short ; the pistil, on the contrary, projects considerably above the corolla. The nectar is not protected by any special arrangement of the flower itself, and is accessible even to very small insects. The stamens ripen before the pistil, and any flying insect, however small, coming from above, would assist in cross-fertilisation. Creeping insects, on the contrary, which in most cases would enter from below, would rob the honey without benefiting the plant. *P. amphibium*, as its name

FIG. 32.—*Linnæa.*

FIG. 33.—*Carlina.*

denotes, grows sometimes in water, sometimes on land. So long, of course, as it grows in water, it is thoroughly protected, and the stem is smooth ; while, on the other hand, those specimens which live on land throw out certain hairs which terminate in sticky glands, and thus prevent small insects from creeping up to the flowers. In this case, therefore, the plant is not sticky, except just when this condition is useful. All these viscous plants, as far as I know, have upright or horizontal flowers.

On the other hand, where the same object is effected by slippery surfaces, the flowers are often pendulous ; creeping creatures being thus kept out of them, just as the pendulous nests of the weaver-bird are a protection from snakes and other enemies. As instances of this kind, I may mention the common Snowdrop and the Cyclamen.

Many flowers close their petals during rain, and this is obviously an advantage, since it prevents the honey and pollen from being spoilt or washed away. I have elsewhere suggested that the so-called "sleep" of flowers has reference to the habits of insects, on the ground that flowers which are fertilised by night-flying insects would derive no advantage from being open in the day ; while, on the other hand, those which are fertilised by bees would gain nothing by being open at night. I confess that I suggested this with much diffidence, but it may now, I think, be regarded as well established.[1]

Silene nutans (Fig. 34), the Nottingham Catchfly, is a very instructive species from this point of view, and indeed illustrates a number of interesting points in the relations between plants and insects. Its life history has recently been well described by Kerner. The upper part of the flowering stem is viscid ; from which it has derived its English name, the Nottingham Catchfly. This prevents the access of ants and other small creeping insects. Each flower lasts three days, or rather three nights. The stamens are ten in number, arranged in two sets, the one set standing in

---

[1] The so-called sleep of leaves is a different problem, and probably enables the plant to support better the cold of night.

front of the sepals, the other in front of the petals.
Like other night flowers, it is white, and opens
towards evening, when it also becomes extremely
fragrant. The first evening, towards dusk, the five
stamens in front of the sepals grow very rapidly for
about two hours, so that they emerge from the
flower; the pollen ripens, and is exposed by the
bursting of the anther. So the flower remains through
the night, very attractive to, and much visited by,
moths. Towards three in the morning the scent

FIG. 34.—*Silene nutans.*

ceases, the anthers begin to shrivel up or drop off, the
filaments turn themselves outwards, so as to be out of
the way, while the petals, on the contrary, begin to
roll themselves up, so that by daylight they close the
aperture of the flower, and present only their brownish-
green undersides to view; which, moreover, are thrown
into numerous wrinkles. Thus, by the morning's
light, the flower has all the appearance of being faded.
It has no smell, and the honey is covered over by
the petals. So it remains all day. Towards evening,
however, everything is changed. The petals unfold

themselves ; by eight o'clock the flower is as fragrant
as before, the second set of stamens have rapidly
grown, their anthers are open, and the pollen again
exposed. By morning the flower is again " asleep,"
the anthers are shrivelled, the scent has ceased, and the
petals rolled up as before. The third evening, again
the same process occurs, but this time it is the pistil
which grows: the long spiral stigmas on the third
evening take the position which on the previous two
had been occupied by the anthers, and can hardly fail
to be dusted by moths with pollen brought from
another flower.

An objection to the view that the sleep of flowers is
regulated by the visits of insects, might be derived
from the cases of those flowers which close early in
the day, the well-known *Tragopogon pratense*, or " John
Go-to-bed at Noon," for instance ; still more, such
species as *Lapsana communis*, or *Crepis pulchra*, which
open before six and close again before ten in the
morning. Bees, however, are very early risers, while
ants come out later, when the dew is off; so that it
might be an advantage to a flower which was quite
unprotected to open early for the bees, and close
again before the ants were out, thus preserving its
honey exclusively for bees.

Thus then I have endeavoured to show in a variety
of cases how beautifully flowers are constructed, so as
to secure their fertilisation by insects. Neither plants
nor insects would be what they are, but for the in-
fluence which each has exercised on the other. Some
plants, indeed, are altogether dependent on insects for
their very existence. We know now, for instance, that

certain plants produce no seeds at all, unless visited by insects. Thus, in some of our colonies, the common Red Clover sets no seeds, on account of the absence of humble bees ; for the proboscis of the hive bee is not long enough to effect the object. According to Mr. Belt, the same is the case, and for the same reason, in Nicaragua, with the scarlet-runner. But even in those instances in which it is not absolutely necessary, it is an advantage that the flowers should be fertilised by pollen brought from a different stock, and with this object in view insects are tempted to visit flowers for the sake of the honey and pollen ; while the colours and scents are useful in making the flowers more easy to find.

Fortunately for us, bees like the same odours as we do ; and as the great majority of flowers are adapted for bees, they are consequently sweet ; but it might have been otherwise, for flies, as already mentioned, prefer unpleasant smells, such as those of decaying meat, and other animal substances on which they live as larvæ, and some flowers, consequently, which are fertilised by them are characterised by very evil odours. Colours also are affected in the same manner, for while bee-flowers (if I may coin such an expression) have generally bright, clear colours, fly-flowers are usually reddish or yellowish brown.

Nevertheless although flowers present us with these beautiful and complex contrivances, whereby the transfer of pollen from flower to flower is provided for, and waste is prevented, yet they appear to be imperfect, or at least not yet perfect in their adaptations. Many small insects obtain access to flowers and rob them of their contents. *Malva rotundifolia* can be, and often is, sucked by bees from the outside, in which case the flower derives no advantage from the visit of the insect. In *Medicago sativa*, also, insects can suck the honey without effecting fertilisation, and the same flower continues to secrete honey after fertilisation has taken place, and when, apparently, it can no longer be of any use. Fritz Müller has observed that though *Posoqueria fragrans* is exclusively fertilised by night-flying insects, many of the flowers open in the day, and consequently remain sterile. It is of course possible that these cases may be explained away; nevertheless, as both insects and flowers are continually altering in their structure, and in their geographical distribution, we should naturally expect to find such instances. Water continually tends to find its own level; animals and plants as constantly tend to adapt themselves to their conditions. For it is obvious that any blossom which differed from the form and size best adapted to secure the due transference of the pollen would be less likely to be fertilised than others; while on the other hand, those richest in honey, sweetest, and most conspicuous, would most surely attract the attention and secure the visits of insects; and thus, just as our gardeners, by selecting seed from the most beautiful varieties, have done so much to adorn our gardens, so have insects, by fertilising the largest and most brilliant flowers, contributed unconsciously, but not less effectually, to the beauty of our woods and fields.[1]

---

[1] I have treated the subject of these chapters at greater length in a little book on Flowers and Insects, forming one of the present series.

FIG. 35.—*Cardamine chenopodifolia.*
a a, ordinary pods ; *b*, subterranean pods.

CHAPTER III.

FRUITS AND SEEDS.

FRUITS and Seeds, though not generally so conspicuous as flowers, are not less interesting.

In considering them, it is fortunately not necessary
to use many technical terms, though it is impossible

to avoid them altogether. In order to understand the
structure of the seed, we must commence with the
flower, to which the seed owes its origin. Now I
have already mentioned, but it may be convenient to
repeat here, if you take such a flower as, say a
Geranium, you will find that it consists of the following
parts: Firstly, there is a whorl of green leaves, known
as the sepals, and together forming the calyx
secondly, a whorl of coloured leaves, or petals,
generally forming the most conspicuous part of the
flower, and called the corolla ; thirdly, a whorl of
organs more or less like pins, which are called stamens,
in the heads, or anthers of which, the pollen is
produced. These anthers are in reality, as Goethe
showed, modified leaves ; in the so-called double
flowers, as, for instance, in our garden roses, they are
developed into coloured leaves like those of the corolla,
and monstrous flowers are not unfrequently met with,
in which the stamens are green leaves, more or less
resembling the ordinary leaves of the plant. Lastly,
in the centre of the flower is the pistil, which also is
theoretically to be considered as constituted of one or
more leaves, each of which is folded on itself, and
called a carpel. Sometimes there is only one carpel.
Generally the carpels have so completely lost the
appearance of leaves, that this explanation of their
true nature requires a considerable amount of faith,
though in others, as for instance in the Columbine
(*Aquilegia*), the original leaf-form can still be traced.
The base of the pistil is the ovary, composed, as I
have just mentioned, of one or more carpels, in which
the seeds are developed. I need hardly say that many

so-called seeds are really fruits ; that is to say, they
are seeds with more or less complex envelopes.

We all know that seeds and fruits differ greatly in
different species.   Some are large, some small ; some
are sweet, some bitter ; some are brightly coloured :
some are good to eat, some poisonous ; some spherical,
some winged, some covered with bristles, some with
hairs ; some are smooth, some very sticky.

We may be sure that there are good reasons for
these differences.   In the case of flowers much light
has been thrown on their various interesting peculiarities by the researches of Sprengel, Darwin, Müller,
and other naturalists.   As regards seeds also, besides
Gærtner's great work, Hildebrand, Krause, Stein-
brinck, Kerner, Grant Allen, Wallace, Darwin, and
others, have published valuable researches, especially
with reference to the hairs and hooks with which so
many seeds are provided, and the other means of
dispersion they possess.   Nobbe also has contributed
an important work on seeds, principally from an agri-
cultural point of view, but the subject as a whole offers
a most promising field for investigation.   It is rather
with a view of suggesting this branch of science to you,
than of attempting to supply the want myself, that I
now propose to call your attention to it.   In doing so
I must, in the first place, express my acknowledg-
ments to Mr. Baker, Mr. Carruthers, Mr. Hemsley,
and especially to Mr. Thiselton Dyer and Sir Joseph
Hooker, for their kind and most valuable assistance.

It is said that one of our best botanists once observed
to another that he never could understand what was
the use of the teeth on the capsules of mosses.   " Oh,"

replied his friend, " I see no difficulty in that, because if it were not for the teeth, how could we distinguish the species ? "

We may, however, no doubt, safely consider that the peculiarities of seeds have reference to the plant itself, and not to the convenience of botanists.

In the first place, then, during growth, seeds in many cases require protection. This is especially the case with those of an albuminous character. It is curious that so many of those which are luscious when ripe, as the Peach, Strawberry, Cherry, Apple, &c., are stringy, and almost inedible, till ripe. Moreover, in these cases, the fleshy portion is not the seed itself, but only the envelope, so that even if the sweet part is eaten the seed itself remains uninjured.

On the other hand, such seeds as the Hazel, Beech, Spanish Chestnut, and innumerable others, are protected by a thick, impervious shell, which is especially developed in many Proteaceæ, the Brazil-nut, the so-called Monkey-pot, the Cocoa-nut, and other palms.

In other cases the envelopes protect the seeds, not only by their thickness and toughness, but also by their bitter taste, as, for instance, in the Walnut. The genus Mucuna, one of the Leguminosæ, is remarkable in having the pods covered with stinging hairs.

In many cases the calyx, which is closed when the flower is in bud, opens when the flower expands, and then after the petals have fallen closes again until the seeds are ripe, when it opens for the second time. This is, for instance, the case with the common Herb Robert (*Geranium robertianum*). In *Atractylis cancellata*, a South European plant, allied to the thistles, the outer envelopes form an exquisite little cage. Another case, perhaps, is that of *Nigella*, the " Devil-in-a-bush," or, as it is sometimes more prettily called, " Love-in-a-mist," of old English gardens.

Again, the protection of the seed is in many cases attained by curious movements of the plant itself. In fact, plants move much more than is generally supposed. So far from being motionless, they may almost be said to be in perpetual movement, though the changes of position are generally so slow that they do not attract attention. This is not, however, always the case. We are all familiar with the Sensitive Plant which droops its leaves when touched. Another species (*Averrhoa bilimbi*) has leaves like those of an Acacia, and all day the leaflets go slowly up and down. *Desmodium gyrans*, a sort of pea living in India, has trifoliate leaves, the lateral leaflets being small and narrow ; and these leaflets, as was first observed by Lady Monson, are perpetually moving round and round, whence the specific name *gyrans*. In these two cases the object of the movement is quite unknown to us. In *Dionæa*, on the other hand, the leaves form a regular fly-trap. Directly an insect alights on them they shut up with a snap.

In a great many cases leaves are said to sleep ; that is to say, at the approach of night they change their position, and sometimes fold themselves up, thus presenting a smaller surface for radiation, and being in consequence less exposed to cold. Mr. Darwin has proved experimentally that leaves which he prevented from moving suffered more from cold than those which were allowed to assume their natural

E

position. He has observed with reference to one
plant, *Maranta arundinacea*, the Arrowroot, a West
Indian species allied to Canna, that if the plant has
had a severe shock it cannot get to sleep for the next
two or three nights.

The sleep of flowers is also probably a case of the
same kind, though, as I have already attempted to
show, it has, I believe, special reference to the visits
of insects ; those flowers which are fertilised by bees,
butterflies, and other day insects, sleep by night,
if at all ; while those which are dependent on moths
rouse themselves towards evening, as already mentioned, and sleep by day. These motions, indeed,
have but an indirect reference to our present subject.
On the other hand, in the Dandelion (*Leontodon*), the
flower-stalk is upright while the flower is expanded,
a period which lasts for three or four days ; it then
lowers itself and lies close to the ground for about
twelve days, while the fruits are ripening, and then
rises again when they are mature. In the Cyclamen
the stalk curls itself up into a beautiful spiral after
the flower has faded.

The flower of the little Linaria of our walls (*L.
cymbalaria*) pushes out into the light and sunshine,
but as soon as it is fertilised it turns round and
endeavours to find some hole or cranny in which it
may remain safely ensconced until the seed is ripe.

In some water-plants the flower expands at the
surface, but after it is faded retreats again to the
bottom. This is the case, for instance, with the
Water Lilies, some species of the Potamogeton, *Trapa
natans.* In Valisneria, again, the female flowers

(Fig. 36, *a*) are borne on long stalks, which reach to the
surface of the water, on which the flowers float.   The
male flowers (Fig. 36, *b*), on the contrary, have short,

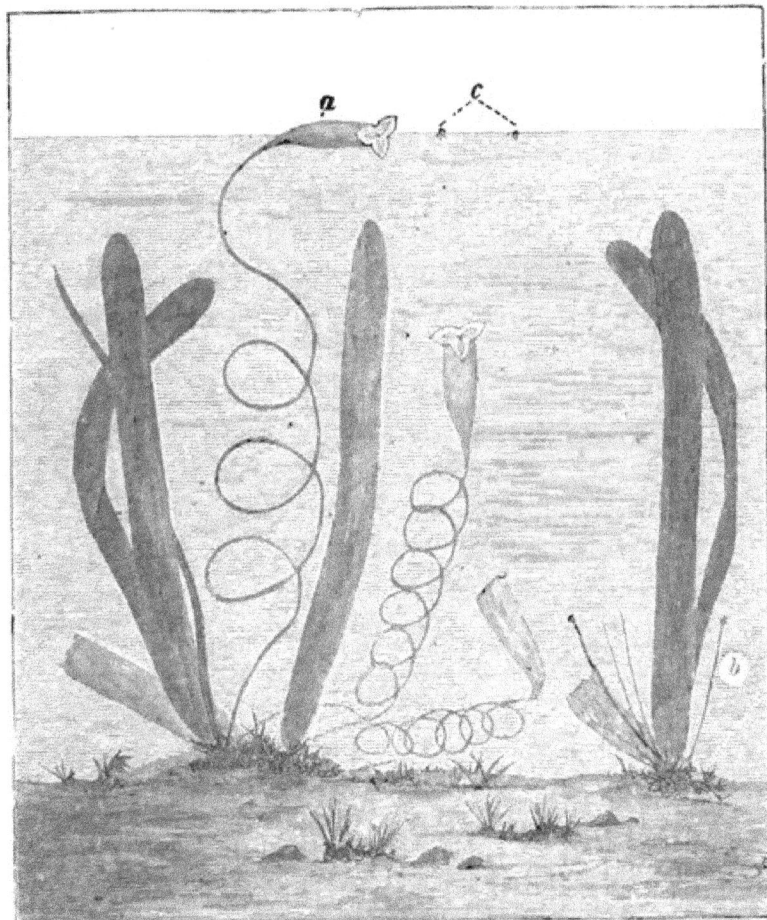

FIG. 36.—*Valisneria spiralis.*
a, female flower ; *b*, male flower ; *c*, floating pollen.

straight stalks, from which, when mature, the pollen
(Fig. 36, *c*) detaches itself, rises to the surface, and,
floating freely on it, is wafted about, so that it comes

in contact with the female flowers.   After fertilisation,
however, the long stalk coils up spirally, and thus
carries the ovary down to the bottom, where the seeds
can ripen in greater safety.

The next points to which I will direct your attention
are the means of dispersion possessed by many seeds.
Farmers have found by experience that it is not
desirable to grow the same crop in the same field
year after year, because the soil becomes more or less
exhausted.   In this respect, therefore, the powers of
dispersion possessed by many seeds are a great
advantage to the species.   Moreover, they are also
advantageous in giving the seed a chance of germi-
nating in new localities suitable to the requirements
of the species.   Thus a common European species,
Xanthium spinosum, has rapidly spread over the whole
of South Africa, the seeds being carried in the wool
of sheep.   From various considerations, however, it
seems probable that in most cases the provision does
not contemplate a dispersion for more than a short
distance.

There are a great many cases in which plants
possess powers of movement directed to the dissemina-
tion of the seed.

I have already referred to the case of the common
Dandelion.   Here the flower-stalk stands more or less
upright while the flower is expanded, a period which
generally lasts for three or four days.   It then lowers
itself, and lies more or less horizontally and concealed
during the time the seeds are maturing, which in our
summers occupies about twelve days.   It then again
rises, and, becoming almost erect, facilitates the

dispersion of the seeds, or, speaking botanically, the fruits, by the wind. Some plants, as we shall see, even sow their seeds in the ground, but these cases will be referred to later on.

In other cases the plant throws its own seeds to some little distance. This is the case with the common *Cardamine hirsuta*, a little plant, I do not like to call it a weed, six or eight inches high, which comes up of itself abundantly on any vacant spot in our kitchen-gardens or shrubberies, and which much resembles that represented in Fig. 17, but without the subterranean pods *b*. The seeds are contained in a pod which consists of three parts, a central membrane, and two lateral walls. When the pod is ripe the walls are in a state of tension. The seeds are loosely attached to the central piece by short stalks. Now, when the proper moment has arrived, the outer walls are kept in place by a delicate membrane, only just strong enough to resist the tension. The least touch, for instance, a puff of wind blowing the plant against a neighbour, detaches the outer wall, which suddenly rolls itself up, generally with such force as to fly from the plant, thus jerking the seeds to a distance of several feet.

In the common Violet, beside the coloured flowers, there are others in which the corolla is either absent or imperfectly developed. The stamens also are small, but contain pollen, though less than in the coloured flowers. In the autumn large numbers of these curious flowers are produced. When very young they look like an ordinary flower-bud (Figs. 37 and 38, *a*), the central part of the flower being entirely covered by the sepals, and the whole having a triangular form. When older (Figs. 37 and 38, *b*) they look at first sight like an ordinary seed capsule, so that the bud seems to pass into the capsule without the flower-stage.

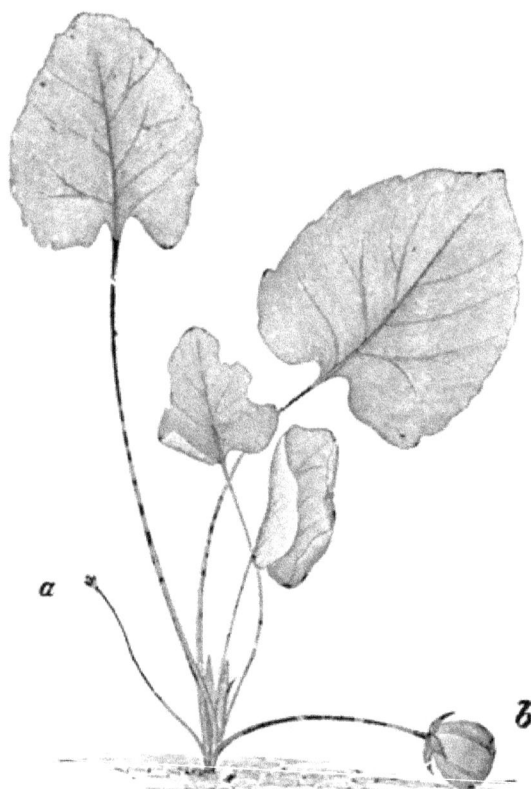

FIG. 37.—*Viola hirta.*
a, young bud ; *b*, ripe seed capsule.

force these capsules into the ground, and thus sow
their own seeds. I have not, however, found this to
be the case, though as the stalk elongates, and the
point of the capsule turns downwards, if the earth be
loose and uneven, it will no doubt sometimes so

Fig. 38.—*Viola canina.*
a, bud ; *b*, bud more advanced ; *c*, capsule open, some of the seeds are already
thrown.

happen. When the seeds are fully ripe, the capsule
opens by three valves and allows them to escape.

In the Dog Violet (*V. canina*, Fig. 38) the case is
very different. The capsules are less fleshy, and,
though pendent when young, at maturity they erect
themselves (Fig. 38, *c*), stand up boldly above the
rest of the plant, and open by the three equal valves

(Fig. 39) resembling an inverted tripod. Each valve contains a row of three, four, or five brown, smooth, pear shaped seeds, slightly flattened at the upper, wider end. Now the two walls of each valve, as they become drier, contract, and thus approach one another, thus tending to squeeze out the seeds. These resist some time, but at length the attachment of the seed to its base gives way, and it is ejected several feet, this being no doubt much facilitated by its form and smoothness. I have known even a gathered specimen

FIG. 39 —*Viola canina.* Seed vessels with seed.

throw a seed nearly 10 feet. Fig. 40 represents a capsule after the seeds have been ejected.

Now we naturally ask ourselves what is the reason for this difference between the species of Violets ; why do *V. odorata* and *V. hirta* conceal their capsules among the moss and leaves on the ground, while *V. canina* and others raise theirs boldly above their heads, and throw the seeds to seek their fortune in the world ? If this arrangement be best for *Viola canina*, why has not *Viola odorata* also adopted it ? The reason is, I believe, to be found in the different mode of growth of these two species. *Viola canina* is a plant with an elongated stalk, and it is easy therefore for the capsule to raise itself above the grass and other low herbage among which violets grow.

V. odorata and *V. hirta*, on the contrary, have, in ordinary parlance, no stalk, and the leaves are radical, *i.e.* rising from the root. This is at least the case in appearance, for, botanically speaking, they rise at the end of a short stalk. Now, under these circumstances,

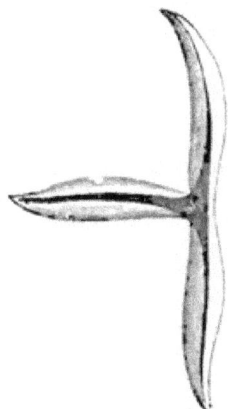

FIG. 40.—*Viola canina.* Seed vessel after ejecting the seed.

if the Sweet Violet attempted to shoot its seeds, the capsules not being sufficiently elevated, the seeds would merely strike against some neighbouring leaf, and immediately fall to the ground. Hence, I think, we see that the arrangement of the capsule in each species is that most suitable to the general habit of the plant.

In the true Geraniums again, as, for instance, in the Herb Robert (Fig. 41), after the flower has faded, the central axis gradually elongates (Fig. 41, *a, c, d*). The

seeds, five in number, are situated at the base of the
column, each being inclosed in a capsule, which

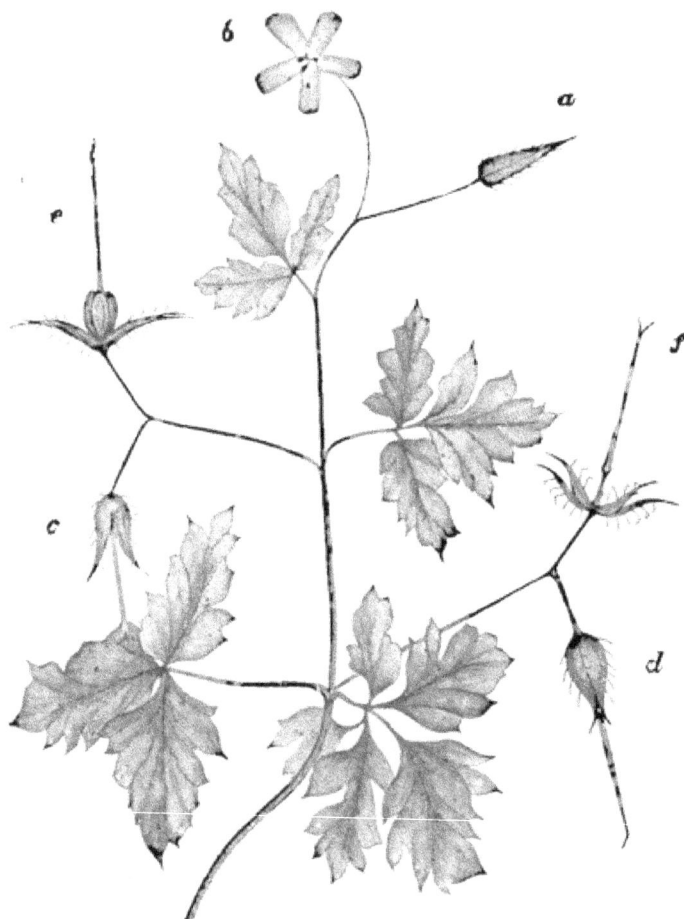

FIG. 41.—HERB ROBERT (*Geranium robertianum*).
a, bud ; *b*, flower ; *c*, flower after the petals have fallen : *d*, flower with seeds nearly ripe ; *e*, flower with ripe seeds ; *f*, flower after throwing seeds.

terminates upwards in a rod-like portion, which at
first forms part of the central axis, but gradually
detaches itself. When the seeds are ripe the ovary

raises itself into an upright position (Fig. 41, *e*) ; the
outer layers of the rod-like termination of the seed-
capsule come to be in a state of great tension, and
eventually detach the rod with a jerk, and thus throw
the seed some little distance.   Fig. 41, *f*, represents
the central rod after the seeds have been thrown.   In

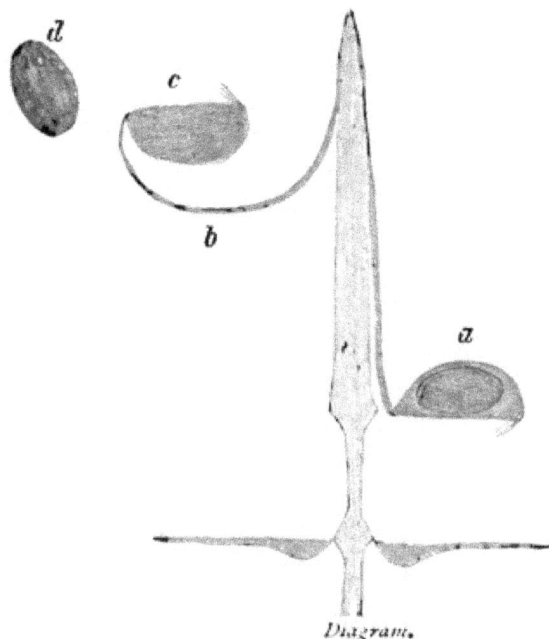

Diagram.

FIG. 42.— *Geranium dissectum.*
a, just before throwing seed ; *b*, just after throwing seed ; *c*, the capsule still attached
to the rod ; *d*, the seed.

some species, as for instance in *Geranium dissectum*,
Fig. 42, the capsule-rod remains attached to the central
column and the seed only is ejected.

It will, however, be remembered that the capsule is,
as already observed, a leaf folded on itself, with the
edges inwards, and in fact in the Geranium the seed-

chamber opens on its inner side. You will, therefore,
naturally observe to me that when the carpel bursts
outwards, the only effect would be that the seed
would be forced against the outer wall of the carpel,
and that it would not be ejected, because the opening
is not on the outer but on the inner side. This
remark is perfectly just, but the difficulty has been
foreseen by our Geraniums, and is overcome by them
in different ways. In some species, as for instance
in *Geranium dissectum,* a short time before the
dehiscence, the seed-chamber places itself at right
angles to the pillar (Fig. 42, *a*). The edges then
separate, but they are provided with a fringe of hairs,
just strong enough to retain the seed in its position,
yet sufficiently elastic to allow it to escape when the
carpels burst away, remaining attached, however, to
the central pillar by their upper ends (Fig. 42, *c*).

In the common Herb Robert (Fig. 43), and some
other species, the arrangement is somewhat different.
In the first place the whole carpel springs away (Fig.
43, *b* and *c*). The seed-chamber (Fig. 43, *c*) detaches
itself from the rod of the carpel (Fig. 43, *b*), and when
the seed is flung away remains attached to it. Under
these circumstances it is unnecessary for the chamber
to raise itself from the central pillar, to which accord-
ingly it remains close until the moment of disruption
(Fig. 41, *e*). The seed-chamber is moreover held in
place by a short tongue which projects a little way
over its base ; while, on the other hand, the lower end
of the rod passes for a short distance between the
seed-capsule and the central pillar. The seed-capsule
has also near its apex a curious tuft of silky hair

(Fig. 43, *c*), the use of which I will not here stop to
discuss.   As the result of all this complex mechanism
the seeds when ripe are flung to a distance which is
surprising when we consider how small the spring is.

In their natural habitat it is almost impossible to
find the seeds when once thrown. I therefore brought

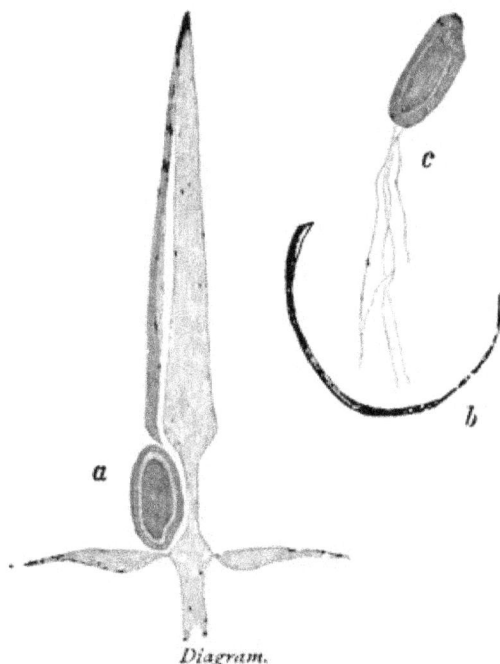

Diagram.
FIG. 43 —*Geranium robertianum.*
a, just before thr.wing the seed ; *b*, the rod ; *c*, the seed inclo-ed in the capsule.

some into the house and placed them on my billiard-
table.   They were thrown from one end completely
beyond the other, in some cases more than twenty feet.

Some species of Vetch, again, and the common
Broom, throw their seeds, owing to the elasticity of
the pods, which, when ripe, open suddenly with a jerk.

Each valve of the pod contains a layer of woody cells,
which, however, do not pass straight up the pod, but
are more or less inclined to its axis (Fig. 44). Consequently, when the pod bursts it does not, as in the case
of Cardamine, roll up like a watch-spring, but twists
itself more or less like a corkscrew.

I have mentioned these species because they are
some of our commonest wild flowers, so that during

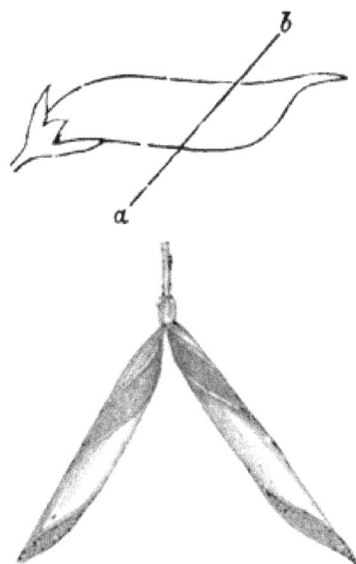

FIG. 44.—COMMON VETCH (*Vicia sepium*).
The line *a b* shows the direction of the woody fibres.

the summer and autumn we may in almost any walk
observe for ourselves this innocent artillery. There
are, however, many other more or less similar cases.
Thus the Squirting Cucumber (*Momordica elaterium*),
a common plant in the south of Europe, and one
grown in some places for medicinal purposes, effects
the same object by a totally different mechanism.

The fruit is a small cucumber (Fig. 45), and when
ripe it becomes so gorged with fluid that it is in a
state of great tension. In this condition a very slight
touch is sufficient to detach it from the stalk, when
the pressure of the walls ejects the contents, throwing
the seed some distance. I have seen them even in
this country sent nearly twenty feet; but in a hotter
climate the plant grows more vigorously, and they

FIG. 45.—THE SQUIRTING CUCUMBER (*Momordica elaterium*).

would doubtless be thrown further. In this case of
course the contents are ejected at the end by which
the cucumber is attached to the stalk. If any one
touches one of these ripe fruits, they are often thrown
with such force as to strike him in the face.

In *Cyclanthera*, a plant allied to the Cucumber,
the fruit is unsymmetrical, one side being round and
hairy, the other nearly flat and smooth. The true apex

of the fruit which bears the remains of the flower, is also
somewhat eccentric, and, when the seeds are ripe, if
it is touched even lightly, the fruit explodes and the
seeds are thrown to some distance.   The mechanism
by which this is effected has been described by
Hildebrand.   The interior of the fruit is occupied by
a loose cellular structure.   The central column, or
placenta, to which the seeds are attached, lies loosely
in this tissue.   Through the solution of its earlier
attachments, when the fruit is ripe, the column adheres
only at the apical end, under the withered remains of
the flower, and at the swollen side.   When the fruit
bursts the placenta unrolls, and thus hurls the seeds to
some distance, being even itself sometimes also torn
away from its attachment.

Other cases of projected seeds are afforded by
Impatiens, Hura, one of the *Euphorbiæ, Collomia,
Oxalis,* some species allied to Acanthus, and by
Arceuthobium, a plant allied to the Mistletoe, and
parasitic on Juniper, which ejects its seeds to a
distance of several feet, throwing them thus from
one tree to another.

Even those species which do not eject their seeds
often have them so placed with reference to the
capsule that they only leave it if swung or jerked by
a high wind.   In the case of trees, even seeds with
no special adaptation for dispersion must in this
manner be often carried to no little distance ; and
to a certain, though less extent, this must hold good
even with herbaceous plants.   It throws light on the,
at first sight, curious fact that in so many plants with
small, heavy seeds, the capsules open not at the

bottom, as one might perhaps have been disposed to expect, but at the top. A good illustration is afforded by the well-known case of the common Poppy (Fig. 46), in which the upper part of the capsule presents a series of little doors (Fig. 46, *a*), through which, when the plant is swung by the wind, the seeds come out one by one. The little doors are

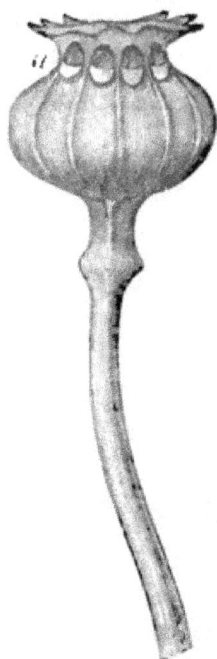

FIG. 46 —SEED-HEAD OF POPPY (*Papaver*).

protected from rain by overhanging eaves, and are even said to shut of themselves in wet weather. The genus Campanula is also interesting from this point of view, because some species have the capsules pendent, some upright, and those which are upright open at the top, while those which are pendent do so at the base.

F

In other cases the dispersion is mainly the work of
the seed itself.   In some of the lower plants, as, for
instance, in many seaweeds, and in some allied fresh-
water plants, such as   *Vaucheria,*   the   spores[1]   are
covered by vibratile cilia, and actually swim about
in the water, like infusoria, till they have found a
suitable spot on which to grow.   Nay, so much do
the spores of some seaweeds resemble animals, that
they are provided with a red " eye-spot" as it has
been called, which, at any rate, seems so far to
deserve the name that it appears to be sensitive to
light.   This mode of progression is, however, only
suitable to water plants.   One group of small, low-
organised   plants,   *Marchantia,*   develop among the
spores a number of cells with spirally thickened walls,
which, by their contractility, are supposed to dis-
seminate the spores.   In the common Horsetails
(*Equisetum*), again, the spores are provided with
curious filaments, terminating in expansions, and
known as " elaters."   These move with great vigour,
and probably serve the same purpose.

In much more numerous cases, seeds are carried by
the wind.   For this of course it is desirable that they
should be light.   Sometimes this object is attained
by the character of the tissues themselves, sometimes
by the presence of empty spaces.   Thus, in *Valerianella
auricula,* the fruit contains three cells, each of which
would naturally be expected to contain a seed.   One
seed only, however, is developed, but, as may be seen
from the figure given in Mr. Bentham's excellent

---

[1] I need hardly observe that, botanically, these are not true seeds,
but rather motile buds.

Handbook of the British Flora, the two cells which contain no seed actually become larger than the one which alone might, at first sight, seem to be normally developed. We may be sure from this that they must be of some use, and, from their lightness, they probably enable the wind to carry the seed to a greater distance than would otherwise be the case.

In other instances the plants themselves, or parts of them, are rolled along the ground by the wind. An example of this is afforded, for instance, by a kind of grass (*Spinifex squarrosus*), in which the mass of inflorescence, forming a large round head, is thus driven for miles over the dry sands of Australia until it comes to a damp place, when it expands and soon strikes root.

So, again, the *Anastatica hierochuntica*, or " Rose of Jericho," a small annual with rounded pods, which frequents sandy places in Egypt, Syria, and Arabia, when dry, curls itself up into a ball or round cushion. and is thus driven about by the wind until it finds a damp place, when it uncurls, the pods open and sow the seeds.

These cases, however, in which seeds are rolled by the wind along the ground, are comparatively rare. There are many more in which seeds are wafted through the air. If you examine the fruit of a Sycamore you will find that it is provided with a wing-like expansion, in consequence of which, if there is any wind when it falls, it is, though rather heavy, blown to some distance from the parent tree. Several cases are shown in Fig. 47 ; for instance, the Maple *a*, Sycamore *b*, Hornbeam *d*, Elm *e*, Birch *f*, Pine *g* :

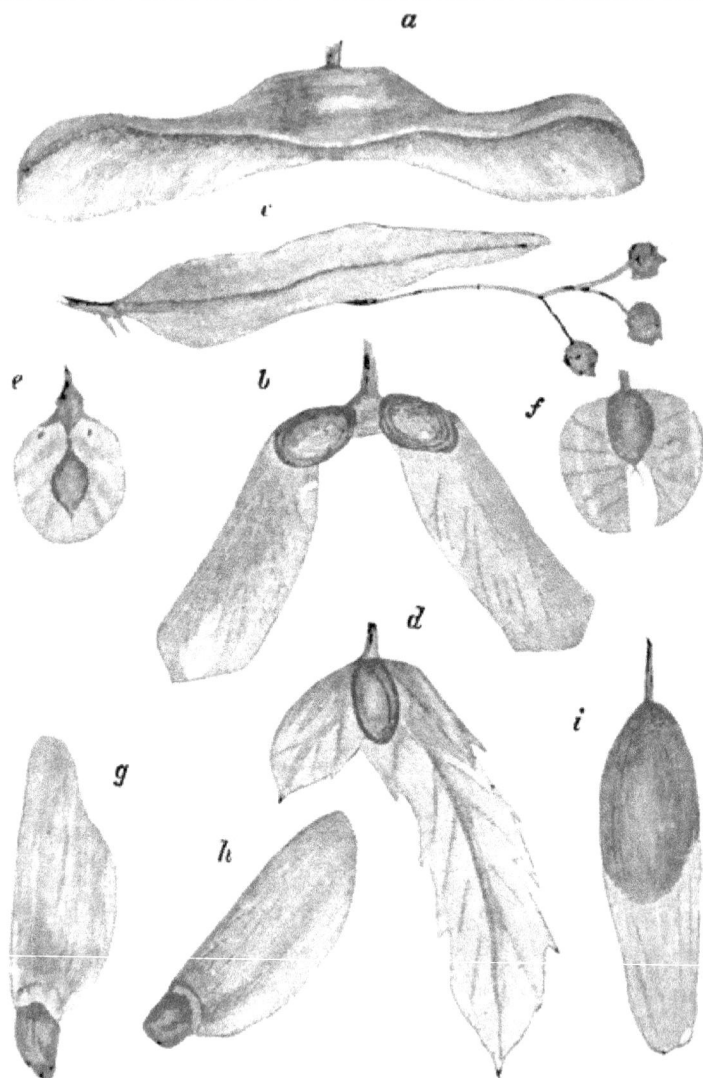

FIG. 47.

a, maple; b, sycamore; c, lime; d, hornbeam; e, elm; f, birch; g, pine; h, fir;
i, ash.

Fir *h*, and Ash *i*, while in the Lime, *c*, the whole
bunch of fruits drop together, and the "bract," as it

is called, or leaf of the flower-stalk, serves the same purpose.

In a great many other plants the same result is obtained by flattened and expanded edges. A beautiful example is afforded by the genus Thysanocarpus, a North American crucifer ; *T. laciniatus* has a distinctly winged pod ; in *T. curvipes* the wings are considerably larger ; lastly, in *T. elegans* and *T. radians* the pods are still further developed in the same direction, *T. radians* having the wing very broad, while in *T. elegans* it has become thinner and thinner in places, until at length it shows a series of perforations. Among our common wild plants we find winged fruits in the Dock (*Rumex*) and in the common Parsnip (*Pastinaca*). But though in these cases the object to be obtained—namely, the dispersion of the seed—is effected in a similar manner, there are differences which might not at first be suspected. Thus in some cases, as, for instance, the Pine, it is the seed itself which is winged ; in *Thlaspi arvense* it is the pod ; in *Entada*, a leguminous plant, the pod breaks up into segments, each of which is winged ; in *Nissolia* the extremity of the pod is expanded into a flattened wing ; lastly, in the Lime, as already mentioned, the fruits drop off in a bunch, and the leaf at the base of the common flower-stalk, or "bract," as it is called, forms the wing.

In *Gouania retinaria* of Rodriguez the same object is effected in another manner ; the cellular tissue of the fruit crumbles and breaks away, leaving only the vascular tissue, which thus forms a net inclosing the seed.

Another mode, which is frequently adopted, is the development of long hairs. Sometimes, as in Clematis, Anemone, and Dryas, these hairs take the form of a long feathery awn. In others the hairs form a tuft or crown, which botanists term a pappus. Of this the Dandelion and John Go-to-bed-at-noon, so called from its habit of shutting its flowers about mid-day, are well-known examples. Tufts of hairs, which are themselves sometimes feathered, are developed in a great many Composites, though some, as, for instance, the Daisy and Lapsana, are without them ; in some very interesting species, of which the common *Thrincia hirta* of our lawns and meadows is one, there are two kinds of fruits, as shown in Fig. 48, *b*, one with a pappus and one without. The former are adapted to seek " fresh woods and pastures new," while the latter stay near the parent plant and perpetuate the race at home.

A more or less similar pappus is found among various English plants—in the Epilobium (Fig. 48, *a*), Thrincia (Fig. 48, *b*), Tamarix (Fig. 48, *c*), Willow Fig. 48, *d*), Cotton Grass (Fig. 48, *e*), and Bulrush (Fig. 48, *f*) ; while in exotic species there are many other cases—as, for instance, the beautiful Oleander. As in the wings, so also in that of the pappus, it is by no means always the same part of the plant which develops into the crown of hairs. Thus in the Valerians and Composites it is the calyx ; in the Bulrush the perianth ; in Epilobium the crown of the seed , in the Cotton Grass it is supposed to represent the perianth ; while in some, as, for instance, in the Cotton plant, the whole outer surface of the seed is

clothed with long hairs.   Sometimes, on the contrary,
the hairs are very much reduced in number, as, for
instance, in some species of *Æschynanthus*, where

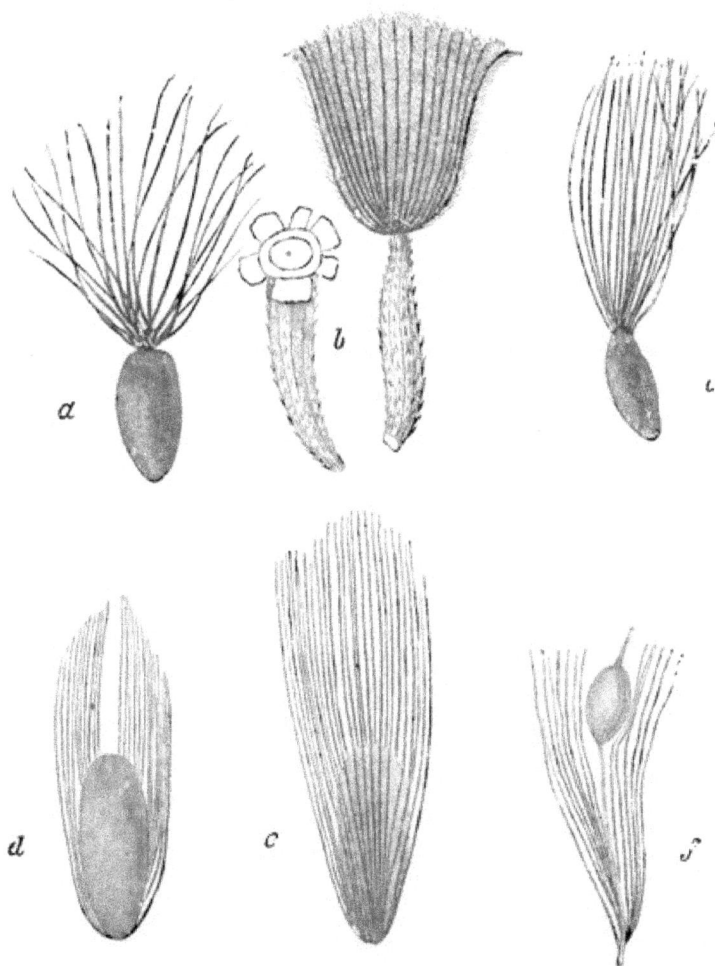

FIG 48.
a, willow herb (*Epilobium*); *b*, two forms of seed of *Thrincia hirta*; *c*, *Tamarix*,
d, willow (*Salix*); *e*, cotton grass (*Eriophorum*); *f*, bulrush (*Typha*).

there are only three, one on one side and two on the
other.   In this case, moreover, the hairs are very

flexible, and wrap round the wool of any animal with which they may come in contact, so that they form a double means of dispersion.

In other cases seeds are wafted by water. Of this the Cocoa-nut is one of the most striking examples. The seeds retain their vitality for a considerable time, and the loose texture of the husk protects them and makes them float. Every one knows that the Cocoa-nut is one of the first plants to make its appearance on coral islands, and it is, I believe, the only palm which is common to both hemispheres.

The seeds of the common Duckweeds (*Lemna*) sink to the bottom of the water in autumn, and remain there throughout the winter ; but in the spring they rise up to the surface again and begin to grow.

Fig. 49.—*Myzodendron*. (After Hooker.)

CHAPTER IV.

In a very large number of cases the diffusion of seeds is effected by animals. To this class belong the fruits and berries. In them an outer fleshy portion becomes pulpy, and generally sweet, inclosing the seeds. It is remarkable that such fruits, in order, doubtless, to attract animals, are, like flowers, brightly coloured—as, for instance, the Cherry, Currant, Apple, Peach, Plum, Strawberry, Raspberry, and many others. This colour, moreover, is not present in the unripe fruit, but is rapidly developed at maturity. In such cases the actual seed is generally protected by a

dense, sometimes almost stony, covering, so that it
escapes digestion, while its germination is perhaps
hastened by the heat of the animal's body. It may
be said that the skin of apple and pear pips is comparatively soft ; but then they are embedded in a
stringy core, which is seldom eaten.

These coloured fruits form a considerable part of the
food of monkeys in the tropical regions of the earth,
and we can, I think, hardly doubt that these animals
are guided by the colours, just as we are, in selecting
the ripe fruit. This has a curious bearing on an
interesting question as to the power of distinguishing
colour possessed by our ancestors in bygone times.
Magnus and Geiger, relying on the well-known fact
that the ancient languages are poor in words for
colour, and that in the oldest books—as, for instance,
in the Vedas, the Zendavesta, the Old Testament,
and the writings of Homer and Hesiod—though
the heavens are referred to over and over again,
its blue colour is never dwelt on, have argued that
the ancients were very deficient in the power of
distinguishing colours, and especially blue. In our
own country Mr. Gladstone has lent the weight of his
great authority to the same conclusion. For my part
I cannot accept this view. There are, it seems to me,
very strong reasons against it, into which I cannot,
of course, now enter ; and though I should rely
mainly on other considerations, the colours of fruits
are not, I think, without significance. If monkeys
and apes could distinguish them, surely we may infer
that even the most savage of men could do so too.
Zeuxis would never have deceived the birds if he had
not had a fair perception of colour.

In these instances of coloured fruits, the fleshy edible part more or less surrounds the true seeds ; in others the actual seeds themselves become edible. In the former the edible part serves as a temptation to animals ; in the latter it is stored up for the use of the plant itself. When, therefore, the seeds themselves are edible they are generally protected by more or less hard or bitter envelopes, for instance the Horse Chestnut, Beech, Spanish Chestnut, Walnut, &c. That these seeds are used as food by squirrels and other animals is, however, by no means necessarily an evil to the plant, for the result is that they are often carried some distance and then dropped, or stored up and forgotten, so that in this way they get carried away from the parent tree.

In another class of instances animals, unconsciously or unwillingly, serve in the dispersion of seeds. These cases may be divided into two classes, those in which the fruits are provided with hooks, and those in which they are sticky. To the first class belong, among our common English plants, the Burdock (*Lappa*, Fig. 50, *a*) ; Agrimony (*Agrimonia*, Fig. 50, *b*) ; the Bur Parsley (*Caucalis*, Fig. 50, *c*) ; Enchanter's Nightshade (*Circæa*, Fig. 50, *d*) ; Goose Grass or Cleavers (*Galium*, Fig. 50, *e*), and some of the Forget-me-nots (*Myosotis*, Fig. 50, *f*). The hooks, moreover, are so arranged as to promote the removal of the fruits. In all these species the hooks, though beautifully formed, are small; but in some foreign species they become truly formidable. Two of the most remarkable are represented on page 77,—*Martynia proboscidea* (Fig. 51, *b*) and *Harpagophyton procumbens* (Fig. 51, *a*). Martynia is a plant of Louisiana, and if

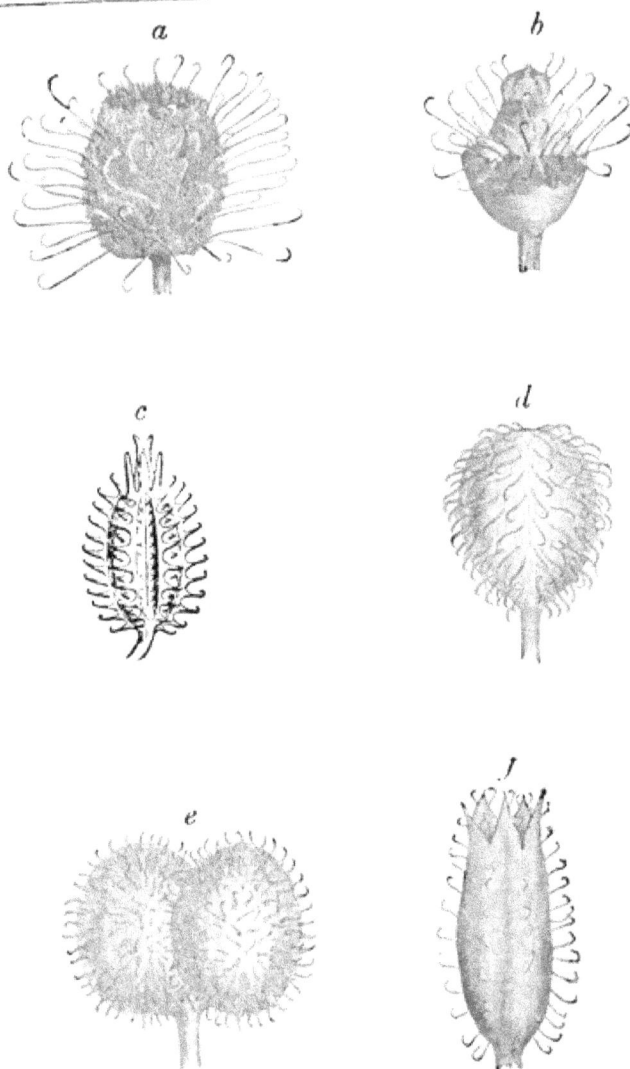

FIG 50

a, burdock (*Lappa*); *b*, agrimony (*Agrimonia*); *c*, bur parsley (*Caucalis*); *d*, enchanter's nightshade (*Circaea*); *e*, cleavers (*Galium*); *f*, forget-me-nots (*Myosotis*)

its fruits once get hold of an animal it is most difficult to remove them.   Harpagophytum is a South African genus.   The fruits are most formidable, and are said

FIG. 51.

a, Harpagophyton procumbens (natural size); *b, Martynia proboscidea* (natural size).

sometimes even to kill lions.   They roll about over the dry plains, and if they attach themselves to the skin,

the wretched animal tries to tear them out, and some
times getting them into his mouth perishes miserably.

The cases in which the diffusion of fruits and seeds
is effected by their being sticky are less numerous, and
we have no well-marked instance among our native
plants.   The common Plumbago of South Europe is
a case which many of you no doubt have observed.
Other genera with the same mode of dispersion are
*Pittosporum, Pisonia, Boerhavia, Siegesbeckia, Grin-
delia, Drymaria,* &c.   There are comparatively few
cases in which the same plant uses more than one of
these modes of promoting the dispersion of its seeds,
still there are some such instances.   Thus in the
common Burdock the seeds have a pappus, while the
whole flower-head is provided with hooks which
readily attach themselves to any passing animal.
Asterothrix, as Hildebrand has pointed out, has three
provisions for dispersion ; it has a hollow appendage,
a pappus, and a rough surface.

But perhaps it will be said that I have picked out
special cases ; that others could have been selected,
which would not bear out, or perhaps would even
negative, the inferences which have been indicated ;
that I have put the cart before the horse ; that the
Ash fruit has not a wing in order that it may be
carried by the wind, or the Burdock hooks that the
heads may be transported by animals, but that
happening to have wings and hooks these seeds are
thus transported.   Now doubtless there are many
points connected with seeds which are still unexplained ; in fact it is because this is so that I was
anxious to direct attention to the subject.   Still I

believe the general explanations which have been given by botanists will stand any test.

Let us take for instance fruits formed on the same type as that of the Ash—that is to say, with a long wing, known to botanists as a Samara. Now such a fruit would be of little use to low herbs, which, however, are so numerous. If the wing was accidental, if it were not developed to serve as a means of dispersion, it would be as likely to occur on low plants and shrubs as on trees. Let us then consider on what kind of plants these fruits are found. They occur on the Ash, Maple, Sycamore, Hornbeam, Pines, Firs and Elm ; while the Lime, as we have seen, has also a leaf attached to the fruits, which answers the same purposes. Seeds of this character therefore occur on a large proportion of our forest trees, and on them alone. But more than this: I have taken one or two of the most accessible works in which seeds are figured, for instance Gærtner's *De Fructibus et Seminibus*, Le Maout and Decaisne (Hooker's translation) *Descriptive and Analytical Botany*, and Baillon's *Histoire des Plantes.* I find thirty genera, belonging to twenty-one different natural orders, figured as having seeds or fruits of this form. They are all trees or climbing shrubs, not one being a low herb.

Let us take another case, that of the plants in which the dispersion of the seeds is effected by means of hooks. Now, if the presence of these hooks were, so to say, accidental, and the dispersion merely a result, we should naturally expect to find some species with hooks in all classes of plants. They would occur, for instance, among trees and on water-plants. On the other hand, if they are developed that they might

adhere to the skin of quadrupeds, then, having
reference to the habits and size of our British
mammals, it would be no advantage for a tree or for
a water-plant to bear hooked seeds. Now, what are
the facts? There are about thirty English species in
which the dispersion of the seeds is effected by means
of hooks, but not one of these is aquatic, nor is one of
them more than four feet high. Nay, I might carry
the thing further. We have a number of minute
plants, which lie below the level at which seeds would
be likely to be entangled in fur. Now none of these,
again, have hooked seeds or fruits. It would also
seem, as Hildebrand has suggested, that in point of
time, also, the appearance of the families of plants in
which the fruits or seeds are provided with hooks
coincided with that of the land Mammalia.

Again, let us look at it from another point of view.
Let us take our common forest-trees, shrubs, and tall
climbing plants ; not, of course, a natural or botanical
group, for they belong to a number of different orders,
but a group characterised by attaining to a height
of say over eight feet. We will in some cases only
count genera ; that is to say, we will count all the wil-
lows, for instance, as one. These trees and shrubs are
plants with which we are all familiar, and are about
thirty-three in number. Now, of these thirty-three
no less than eighteen have edible fruits or seeds, such
as the Plum, Apple, Arbutus, Holly, Hazel, Beech,
and Rose. Three have seeds which are provided with
feathery hairs ; and all the rest, namely, the Lime,
Maple, Ash, Sycamore, Elm, Hop, Birch, Hornbeam,
Pine, and Fir are provided with a wing. Moreover,
as will be seen by the following table, the lower trees

and shrubs, such as the Cornel, Guelder Rose, Rose, Thorn, Privet, Elder, Yew, and Holly have generally edible berries, much eaten by birds. The winged seeds or fruits characterise the great forest trees.

TREES, SHRUBS, AND CLIMBING SHRUBS NATIVE OR NATURALISED IN BRITAIN.

	Seed or Fruit.			
	Edible.	Hairy.	Winged.	Hooked.
Clematis vitalba		×		
Berberis vulgaris	×			
Lime (*Tilia Europæa*) . .			×	
Maple (*Acer*)			×	
Spindle Tree (*Euonymus*) .	×.			
Buckthorn (*Rhamnus*) . .	×			
Sloe (*Prunus*)	×			
Rose (*Rosa*)	×			
Apple (*Pyrus*)	×			
Hawthorn (*Cratægus*) . . .	×			
Medlar (*Mespilus*)	×			
Ivy (*Hedera*)	×			
Cornel (*Cornus*)	×			
Elder (*Sambucus*)	×			
Guelder Rose (*Viburnum*) .	×			
Honeysuckle (*Lonicera*) . .	×			
Arbutus (*Arbutus*)	×			
Holly (*Ilex*)	×			
Ash (*Fraxinus*)			×	
Privet (*Ligustrum*)	×			
Elm (*Ulmus*)			×	
Hop (*Humulus*)			×	
Alder (*Alnus*)[1]				
Birch (*Betula*)			×	
Hornbeam (*Carpinus*) . . .			×	
Nut (*Corylus*)	×			
Beech (*Fagus*)	×			
Oak (*Quercus*)	×			
Willow (*Salix*)		×		
Poplar (*Populus*)		×		
Pine (*Pinus*)			×	
Fir (*Abies*)			×	
Yew (*Taxus*)	×			

[1] S.me species of Alder have winged fruits.

Or let us take one natural order. That of the
Roses is particularly interesting. In the genus *Geum*
the fruit is provided with hooks; in *Dryas* it
terminates in a long feathered awn, like that of
Clematis. On the other hand, several genera have
edible fruits; but it is curious that the part of a plant
which becomes fleshy, and thus tempting to animals,
differs considerably in the different genera. In the
Blackberry, for instance, and in the Raspberry, the
carpels constitute the edible portion. When we eat a
Raspberry we strip them off and leave the receptacle
behind; while in the Strawberry the receptacle constitutes the edible portion; the carpels are small,
hard, and closely surround the seeds. In these genera
the sepals are situated below the fruit. In the Rose
on the contrary, it is the peduncle that is swollen and
inverted, so as to form a hollow cup, in the interior of
which the carpels are situated. Here we must remember that the sepals are situated above, not below,
the fruit. Again, in the Pear and Apple, it is the
ovary which constitutes the edible part of the fruit,
and in which the pips are embedded. At first sight
the fruit of the Mulberry—which, however, belongs to
a different family—closely resembles that of the
Blackberry. In the Mulberry, however, it is the
sepals which become fleshy and sweet.

The next point is that seeds should be in a spot
suitable for their growth. In most cases, the seed
lies on the ground, into which it then pushes its little
rootlet. In plants, however, which live on trees, the
case is not so simple, and we meet some curious
contrivances. Thus, the Mistletoe, as we all know, is

parasitic on trees. The fruits are eaten by birds, and the droppings often therefore fall on the boughs ; but if the seed was like that of most other plants it would soon fall to the ground, and consequently perish. Almost alone among those of English plants it is extremely sticky, and thus adheres to the bark.

I have already alluded to an allied genus, *Arceuthobium*, parasitic on Junipers, which throws its seeds to a distance of several feet. These also are very viscid, or, to speak more correctly, are embedded in a very viscid mucilage, so that if they come in contact with the bark of a neighbouring tree they stick to it.

Dr. Watt has described a curious peculiarity in another species of the same family. The fruit, like that of the Mistletoe and of most other species of this order, consists of a mass of viscid pulp surrounding a single seed, and when detached from the parent plant it adheres to whatever it may fall on. There it germinates. The radicle when it has grown to about an inch in length develops on its extremity a flattened disc, and then curves about until the disc is applied to some neighbouring object. If the spot to which the disc has fastened is suitable, the development of the plant proceeds there. If on the contrary the spot be not suitable, the radicle straightens itself, tears the viscid berry away from whatever it has adhered to, and raises it in the air. The radicle then again curves, and the berry is carried by it to another spot where it again adheres. The disc then detaches itself, and by the curving of the radicle is advanced to another spot where it again fixes itself. Dr. Watt says he has seen

this happen several times, and thus the young plant
seems to select certain places in preference to others.
They have been observed for instance to quit the
leaves, on which they must often alight, and move on
to the stem.

Another very interesting genus, again of the same
family, is *Myzodendron* (Fig. 49), a Fuegian species, described by Sir Joseph Hooker, and parasitic on the Beech.
Here the seed is not sticky, but is provided with four
flattened flexible appendages. These catch the wind
and thus carry the seed from one tree to another. As
soon, however, as they touch any little bough the arms
twist round it and there anchor the seed.

In many epiphytes the seeds are extremely numerous and minute. Their great numbers increase the
chance that the wind may waft some of them to the
trees on which they grow ; and as they are then fully
supplied with nourishment they do not require to
carry any store with them. Moreover their minute
size is an advantage, as they are carried into any
little chink or cranny in the bark ; while a larger or
heavier seed, even if borne against a suitable tree,
would be more likely to drop off. In the genus
Neumannia, the small seed is produced at each end
into a long filament which must materially increase its
chance of adhering.

Among terrestrial species there are not a few
cases in which plants are not contented simply to
leave their seeds on the surface of the soil, but actually
sow them in the ground.

Thus in *Trifolium subterraneum*, one of our rarer
English Clovers, only a few of the florets become

perfect flowers ; the others form a rigid pointed head which at first is turned upwards, and as their ends are close together, constitute a sort of spike. At first, I say, the flower-heads point upwards like those of other Clovers, but as soon as the florets are fertilised, the flower-stalks bend over and grow downwards, forcing the flower-head into the ground, an operation much facilitated by the peculiar construction and arrangement of the imperfect florets. The florets are, as Darwin has shown, no mere passive instruments. So soon as the flower-head is in the ground they begin, commencing from the outside, to bend themselves towards the peduncle, the result of which of course is to drag the flower-head further and further into the ground. In most Clovers each floret produces a little pod. This would in the present species be useless, or even injurious ; many young plants growing. in one place would jostle and starve one another Hence we see another obvious advantage in the fact that only a few florets perfect their seeds.

I have already alluded to our Cardamines, the pods of which open elastically and throw their seeds some distance. A Brazilian species, *C. chenopodifolia*, Fig. 35, p. 45, besides the usual long pods, Fig. 35, *a a*, produces also short pointed ones, Fig. 35, *b b*, which it buries in the ground.

Arachis hypogæa is the ground-nut of the West Indies. The flower is yellow and resembles that of a pea, but has an elongated calyx, at the base of which, close to the stem, is the ovary. After the flower has faded, the young pod, which is oval, pointed, and very minute, is carried forward by the growth of the stalk, which becomes several inches long and curves downwards so as generally to force the pod into the ground. If it fails in this, the pod does not develop,

FIG. 52.—*Vicia amphicarpa.*
a a, ordinary pods; *b b,* subterranean pods.

but soon perishes; on the other hand, as soon as it is underground the pod begins to grow and develops two large seeds.

In *Vicia amphicarpa,* Fig. 52, a South European

species of Vetch, there are two kinds of pods. One
of the ordinary form and habit (*a*), the other (*b*) oval,

FIG 53.—*Lathyrus amphicarpos.* (After Sowerby.)
a, ordinary pods ; *b*, subterranean pods.

pale, containing only two seeds borne on underground
stems, and produced by flowers which have no corolla.
   Again, a species of the allied genus *Lathyrus*,
Fig. 53, *L. amphicarpos*, affords us another case of the
same phenomenon.

Other species possessing the same faculty of burying their seeds are *Okenia hypogæa*, several species
of *Commelyna*, and of *Amphicarpæa, Voandzeia sub-
terranea, Scrophularia arguta*, &c. ; and it is very
remarkable that these species are by no means nearly
related, but belong to distinct families, namely, the
Cruciferæ, Leguminosæ, Commelynaceæ, Violarieæ, and
Scrophulariaceæ.

Moreover it is interesting that in *L. amphicarpos*, as
in *Vicia amphicarpa* and *Cardamine chenopodifolia*, the
subterranean pods differ from the usual and aërial form
in being shorter and containing fewer seeds. The reason
of this, is, I think, obvious.   In the ordinary pods the
number of seeds of course increases the chance that
some will find a suitable place.   On the other hand
the subterranean ones are carefully sown, as it were,
by the plant itself.   Several seeds together would
only jostle one another, and it is therefore better that
one or two only should be produced.

In the *Erodiums*, or Crane's Bills, the fruit is a
capsule which opens elastically, in some species
throwing the seeds to some little distance.   The seeds
themselves are more or less spindle-shaped, hairy
and produced into a twisted hairy awn as shown in
Fig. 54, representing a seed of *E. glaucophyllum*.
The number of spiral turns in the awn depends upon
the amount of moisture ; and the seed may thus be
made into a very delicate hygrometer, for if it be
fixed in an upright position, the awn twists or un-
twists according to the degree of moisture, and its
extremity thus may be so arranged as to move up and
down like a needle on a register. It is also affected by

heat.   Now if the awn were fixed instead of the seed,
it is obvious that during the process of untwisting, the
seed itself would be pressed downwards, and, as M.
Roux has shown, this mechanism thus serves actually
to bury the seed.   His observations were made on an
allied species, *Erodium cincoium*, which he chose on
account of its size.   He found that
if a seed of this plant is laid on the
ground, it remains quiet as long as it
is dry ; but as soon as it is moistened
—*i.e.* as soon as the earth becomes
in a condition to permit growth—
the outer side of the awn contracts,
and the hairs surrounding the seed
commence to move outwards, the
result of which is gradually to raise
the seed into an upright position with
its point on the soil.   The awn then
commences to unroll and consequently
to elongate itself upwards, and he
suggests that as it is covered with
reversed hairs, it will probably press
against some blade of grass or other
obstacle, which will prevent its moving
up, and will therefore tend to drive

Fig. 14.—*Erodium glaucophyllum.* (After Sweet.)

the seed into the ground.   If then
the air becomes drier, the awn will again roll up,
in which action M. Roux thought it would tend to
draw up the seed, but from the position of the hairs
the feathery awn can easily slip downwards, and
would therefore not affect the seed.   When moistened
once more, it would again force the seed further

downwards, and so on until the proper depth was obtained. A species of Anemone (*A. montana*) again has essentially the same arrangement, though belonging to a widely separated order.

A still more remarkable instance is afforded by a beautiful South European grass, *Stipa pennata* (Fig. 55), the structure of which has been described by Vaucher, and more recently, as well as more completely, by Frank Darwin. The actual seed is small, with a sharp point, and stiff, short hairs pointing backwards. The upper end of the seed is produced into a fine twisted corkscrew-like rod, which is followed by a plain cylindrical portion, attached at an angle to the corkscrew, and ending in a long and beautiful feather, the whole being more than a foot in length. The long feather, no doubt, facilitates the dispersion of the seeds by wind ; eventually, however, they sink to the ground, which they tend to reach, the seed being the heaviest portion, point downwards. So the seed remains as long as it is dry, but if a shower comes on, or when the dew falls, the spiral unwinds, and if, as is most probable, the surrounding herbage or any other obstacle prevents the feathers from rising, the seed itself is forced down and so driven by degrees into the ground.

I have already mentioned several cases in which plants produce two kinds of seeds, or at least of pods, the one being adapted to burying itself in the ground. Heterocarpism, if I may term it so, or the power of producing two kinds of reproductive bodies, is not confined to these species. There is, for instance, a North African species of Corydalis (*C. heterocarpa* of

FIG. 55.—Seed of *Stipa pennata.* (Natural size.)

Durieu) which produces two kinds of seed (Fig. 56), one somewhat flattened, short and broad, with rounded angles ; the other elongated, hooked, and shaped like a shepherd's crook with a thickened staff.   In this case the hook in the latter form perhaps serves for dispersion.

Our common *Thrincia hirta* (Fig. 48*b*) also possesses, besides the fruits with the well-known feathery crown, others which are destitute of such a provision, and which probably therefore are intended to take root at home.

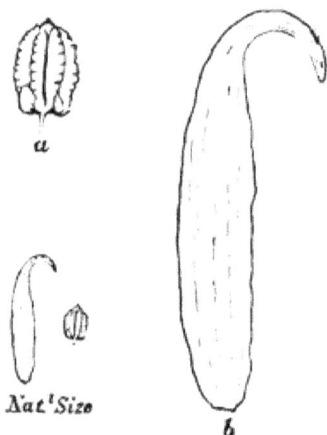

FIG. 56.—Seeds of *Corydalis heterocarpa.*

Mr. Drummond, in the volume of *Hooker's Journal of Botany* for 1842, has described a species of *Alismaceæ* which has two sorts of seed-vessels ; the one produced from large floating flowers, the other at the end of short submerged stalks.   He does not, however, describe either the seeds or seed-vessels in detail.

Before concluding I will say a few words as to the very curious forms presented by certain seeds and fruits. The pods of Lotus, for instance, quaintly resemble a bird's foot, even to the toes; whence the specific name of one species, ornithopodioides; those of Hippocrepis remind one of a horse-shoe; those of Trapa bicornis have an absurd resemblance to the skeleton of a bull's head. These likenesses appear to be accidental, but there are some which probably are of use to the plant. For instance there are two species of Scorpiurus, Fig. 57, the pods of which lie on the ground, and so curiously resemble the one (*S. subvillosa*, Fig. 57, *a*) a centipede, the other (*S. vermiculata*, Fig. 57, *b*) a worm or caterpillar, that it is almost impossible not to suppose that the likeness must be of some use to the plant. May it not be possible that in these cases birds carry the seeds some little distance before they find out that they are not really insects?

The pod of *Biserrula pelecinus* (Fig. 58) also has a striking resemblance to a flattened centipede; while the seeds of *Abrus precatorius*, both in size and in their very striking colour, mimic a small beetle, *Artemis circumusta*.

Mr. Moore has recently called attention to other cases of this kind. Thus the seed of *Martynia diandra* much resembles a beetle with long antennæ; several species of Lupins have seeds much like spiders, and those of *Dimorphochlamys*, a gourdlike plant, mimic a piece of dry twig. In the common Castor Oil plants (Fig. 58*a*), though the resemblance is not so close, still at a first glance the seeds might readily be taken for beetles or ticks. In many Euphorbiaceous plants, as for instance in Jatropha (Fig. 58*b*),

the resemblance is even more striking. The seeds
have a central line resembling the space between the

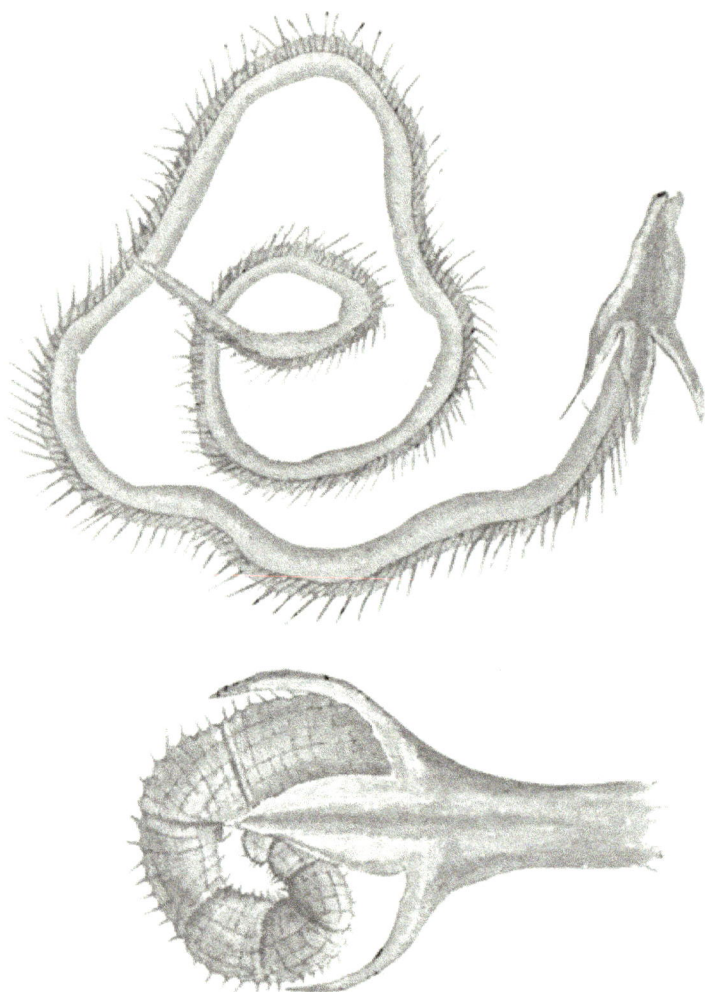

FIG. 57.
a. pod of *Scorpiurus sulcata; b, Scorpiurus vermiculata.*

elytra, dividing and slightly diverging at the end,
while between them the end of the abdomen seems

to peep ; at the anterior end the seeds possess a small
lobe, or caruncle, which mimics the head or thorax of
the insect, and which even seems specially arranged
for this purpose ; at least it would seem from experiments made at Kew that the carunculus exercises
no appreciable effect during germination. In *Tri-
chosanthes anguina* the long pods hang down, and
alike in size, form, colour and attitude closely resemble
snakes, as the specific name denotes.

These resemblances might benefit the plant in one
of two ways. If it be an advantage to the plant

FIG. 58.—Pod of   FIG. 59*a*.—Seed of   FIG. 59*b*—Seed of
Biserrula.   Castor Oil (*Ricinus*).   *Jatropha.*

that the seeds should be swallowed by birds, their
resemblance to insects might lead to this result. On
the other hand if it be desirable to escape from
graminivorous birds, then the resemblance to insects
would serve as a protection. We do not, however,
yet know enough about the habits of these plants
to solve this question.

Indeed, as we have gone on, many other questions
will, I doubt not, have occurred to you, which we are
not yet in a position to answer. Seeds, for instance,
differ almost infinitely in the sculpturing of their

surface. But I shall wofully have failed in my object if I have given the impression that we know all about seeds. On the contrary, there is not a fruit or a seed, even of one of our commonest plants, which would not amply justify and richly reward the most careful study.

In this, as in other branches of science, we have but made a beginning. We have learnt just enough to perceive how little we know. Our great masters in natural history have immortalised themselves by their discoveries, but they have not exhausted the field ; and if seeds and fruits cannot vie with flowers in the brilliance and colour with which they decorate our gardens and our fields, still they surely rival, it would be impossible to excel, them, in the almost infinite variety of the problems they present to us, the ingenuity, the interest, and the charm of the beautiful contrivances which they offer for our study and our admiration.

FIG. 39.—The Beech.

CHAPTER V.

LEAVES.

MR. RUSKIN, in one of his most exquisite passages, has told us that "Flowers seem intended for the solace of ordinary humanity : children love them ; tender, contented, ordinary people love them. They are the cottager's treasure ; and in the crowded town mark, as with a little broken fragment of rainbow, the windows of the workers in whose heart rests the covenant of peace." I should be ungrateful indeed

did I not fully feel the force of this truth ; but
it will be admitted that the beauty of our woods
and fields is due at least as much to foliage as to
flowers.

In the words of the same author, " The leaves of
the herbage at our feet take all kinds of strange
shapes, as if to invite us to examine them. Star-
shaped, heart-shaped, spear-shaped, arrow-shaped,
fretted, fringed, cleft, furrowed, serrated, sinuated, in
whorls, in tufts, in spires, in wreaths, endlessly ex-
pressive, deceptive, fantastic, never the same from
footstalk to blossom, they seem perpetually to tempt
our watchfulness and take delight in outstripping our
wonder."

Now, why is this marvellous variety, this inex-
haustible treasury of beautiful forms ?   Does it result
from some innate tendency of each species ?   Is it
intentionally designed to delight the eye of man ? or
has the form, and size, and texture, some reference to
the structure and organization, the habits and requirements, of the whole plant ?

I do not propose now to discuss any of the more
unusual and abnormal forms of leaves : the pitchers
of Nepenthes or Cephalotus, the pitfalls of Sarracenia
or Darlingtonia, the spring-trap leaves of Dionæa,
the scarcely less effective though less striking contrivances in our own Drosera or Pinguicula, nor the
remarkable power of movement which many leaves
present, whether in response to an external stimulus,
as in certain species of Mimosa, Oxalis, &c., or as a
spontaneous periodic movement, such as the "sleep"

of many leaves, or the nearly continuous rotation of the lateral leaflets of Desmodium. I propose, rather, to ask you to consider the structure, and especially the forms, of the common every-day leaves of our woods and fields.

In talking the subject over with friends, I have found a widely prevalent idea that the beauty and variety of leaves are a beneficent arrangement made specially with reference to the enjoyment and delight of man. I have, again, frequently been met by the opinion that there is some special form, size, and texture of leaf inherently characteristic of each species; that the cellular tissue tends to " crystallize," as it were, into some particular form, quite irrespective of any advantage to the plant itself.

Neither of these views will, I think, stand the test of careful examination.

In the first place, let us consider the size of the leaf. On what does this depend ? In herbs we very often see that the leaves decrease towards the end of the shoot, while in trees the leaves, though not identical, are much more uniform in size.

Again, if we take a twig of Hornbeam, we shall find that the six terminal leaves have together an area of about 14 square inches, and the section of the twig has a diameter of ·06 of an inch. In the beech the leaves are rather larger, six of them having an area of perhaps 18 inches, and, corresponding with this greater leaf-surface, we find that the twig is somewhat stouter, say ·09 of an inch. Following this up, we shall find that, *cæteris paribus*, the size of the leaf has

H 2

relation to the thickness of the stem. This is clearly
shown in the following table :—

Impression of Stalk below the Sixth Leaf.

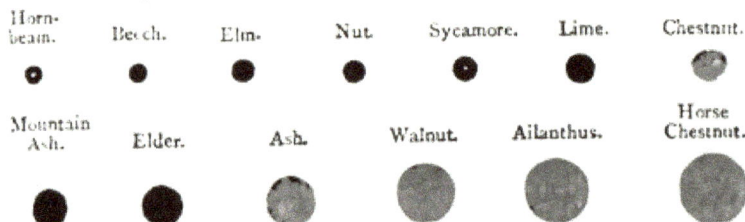

Horn-beam.	Beech.	Elm.	Nut	Sycamore.	Lime.	Chestnut.
●	●	●	●	●	●	●

Mountain Ash.	Elder.	Ash.	Walnut.	Ailanthus.	Horse Chestnut.
●	●	●	●	●	●

	Diameter of Stem in inches.		Approximate Area of Six Upper Leaves in inches.
Hornbeam	·06	...	14
Beech . . .	·09	..	18
Elm	·11	...	34
Nut	·13	...	55
Sycamore	·13	...	60
Lime	·14	...	60
Chestnut	·15	...	72
Mountain Ash	·16	...	60
Elder	·18	...	93
Ash	·18	...	100
Walnut	·25	...	220
Ailanthus	·3	...	240
Horse Chestnut	·3	...	300

In the Elm the numbers are ·11 and 34, in the
Chestnut ·15 and 72, and in the Horse Chestnut the
stem has a thickness of ·3, and the six leaves have
an area often of 300 square inches. Of course, however, these numbers are only approximate. Many
things have to be taken into consideration. Strength,
for instance, is an important element. Thus the
Ailanthus, with a stem equal in thickness to that of

the Horse Chestnut, carries a smaller area of leaves, perhaps because it is less compact. Again, the weight of the leaves must doubtless be taken into consideration. Thus in some sprays of Ash and Elder of equal diameter, which I examined, the former bore the larger expanse of leaves; not only, however, is the stem of the Elder less compact, but the Elder leaves, though not so large, were quite as heavy, if not indeed a little heavier. I was for some time puzzled by the fact that, while the terminal shoot of the Spruce is somewhat thicker than that of the Scotch Fir, the leaves are not much more than $\frac{1}{2}$ as long. May this not perhaps be due to the fact that they remain on the tree more than twice as long, so that the total leaf area borne by the branch is greater, though the individual leaves are shorter? Again, it will be observed that the leaf area of the Mountain Ash is small compared to the stem, and it may, perhaps, not be unreasonable to suggest that this may be connected with the habit of the tree to grow in bleak and exposed situations. The position of the leaves, the direction of the bough, and many other elements would have also to be taken into consideration, but still it seems clear that there is a correspondence between thickness of stem and size of leaf. This ratio, moreover, when taken in relation with the other conditions of the problem, has, as we shall see, a considerable bearing not only on the size, but also on the form of the leaf.

The Mountain Ash has been a great puzzle to me; it is, of course, a true Pyrus, and is merely called Ash from the resemblance of its leaves to those of the common Ash. But the ordinary leaves of a Pear are, as we all know, simple and ovate, or obovate. Why, then, should those of the Mountain Ash be so entirely different? May, perhaps, some light be thrown on this by the arrangement of the leaves? They are situated some distance apart, and though, as shown in the table, they are small in comparison to the diameter of the stem, still they attain a size of 15 square inches, or even more. Now, if they were of the same form as the ordinary Pear leaf, they would be about 7 inches long by 2 to 3 in breadth. The Mountain Ash, as we know, lives in mountainous and exposed localities, and such a leaf would be unsuitable to withstand the force of the wind in such situations. From this point of view, the division into leaflets seems a manifest advantage.

Perhaps it will be said that in some trees the leaves are much more uniform in size than in others. This is true. The Sycamore, for instance, varies greatly; in the specimen tabulated, the stem was ·13 in diameter, and the area of the six upper leaves was 60 square inches. In another, the six upper leaves had an area of rather over 100 inches, but in this case the diameter of the stem was ·18.

Another point is the length of the internode. In such trees as the Beech, Elm, Hornbeam, &c., the distance from bud to bud varies comparatively little, and bears a tolerably close relation to the size of the leaf. In the Sycamore, Maple, &c., on the contrary, the length varies greatly.

Now, if, instead of looking merely at a single leaf, we consider the whole bough of any tree, we

shall, I think, see the reason of their differences of
form.

Let us begin, for instance, with the common Lime
(Fig. 60). The leaf-stalks are arranged at an angle of
about 40° with the branch, and the upper surfaces of
the leaves are in the same plane with it. The result
is, they are admirably adapted to secure the maxi-
mum of light and air. Let us take, for instance, the
second or third leaf in Fig. 60. They are $4\frac{1}{2}$ inches long

Fig. 60.—Lime.

and very nearly as broad. The distance between the
two leaves on each side is also just $4\frac{1}{2}$ inches, so that
they exactly fill up the interval. In *Tilia parvifolia*
the arrangement is similar, but leaves and internodes are both less, the leaves, say $1\frac{1}{2}$ inch, and the
internodes ·6.

In the Beech, the general plane of the leaves is
again that of the branch (Fig. 61), but the leaves
themselves are ovate in form, and smaller, being only

from 2 to 3 inches in length. On the other hand, the
distance between the internodes is also smaller, being,
say, $1\frac{1}{4}$ inch against something less than 2 inches.
The diminution in length of the internode is not,
indeed, exactly in proportion to that of the leaf, but,
on the other hand, the leaf does not make so wide an
angle with the stem. To this position is probably
due the difference of form. The outline of the basal
half of the leaf fits neatly to the branch, that of the
upper half follows the edge of the leaf beyond, and

FIG. 61.—Beech.

the form of the inner edge being thus determined,
decides the outer one also.

In the Nut (*Corylus*), the internodes are longer
and the leaves correspondingly broader. In the Elm
the ordinary branches have leaves resembling, though
rather larger than, those of the Beech; but in vigorous
shoots (*Ulmus*, Fig. 62), the internodes become longer
and the leaves correspondingly broader and larger,
so that they come nearly to resemble those of the
Nut.

But it may be said that the Spanish Chestnut

(*Castanea vulgaris*, Fig. 63) also has alternate leaves
in a plane parallel to that of the branch, and with internodes of very nearly the same length as the Beech.
That is true; but, on the other hand, the terminal
branches of the Spanish Chestnut are stouter in proportion. Thus, immediately below the sixth leaf, the
Chestnut stalk may be ·15 of an inch in thickness, that

FIG. 62.—Elm

of the Beech not much more than half as much.
Consequently, the Chestnut could, of course supposing
the strength of the wood to be equal, bear a greater
weight of leaf; but, the width of the leaf being determined by the distance between the internodes, the
leaf is, so to say, compelled to draw itself out. In
Fig. 64 I have endeavoured to illustrate this by placing
a spray of Beech over one of Spanish Chestnut.

Moreover, not only do the leaves on a single twig thus admirably fit in with one another, but they are also adapted to the ramification of the twigs themselves. Fig. 59 shows a bough of Beech seen from above, and it will be observed that the form of the leaves is such that, while but little space is lost, there is scarcely any over-lapping.   Each fits in perfectly with the rest.

Fig. 63.—Castanea.   Fig. 64.—Castanea and Beech.

The leaves of the Yew (Fig. 65) belong to a type very different from those which we have hitherto been considering.   They are long, narrow, and arranged all round the stem, but spread right and left, so that they lie in one plane, parallel to the direction of the branchlet, and their width bears just such a relation to their distance apart that when so spread out their edges almost touch.

The leaves of Conifers are generally narrow and needle-like. I would venture to suggest that this may be connected with the greater uniformity in the structure of the wood as compared with that of Dicoty-

ledons, such as the Beech, Oak, &c. The leaves of
the Scotch Pine (*Pinus sylvestris*) are needle-like,
$1\frac{1}{2}$ inch in length and $\frac{1}{20}$ in diameter. They are
arranged in pairs, each pair enclosed at the base in a
sheath. One inch of stem bears about fifteen pairs
of leaves. Given this number of leaves in such a
space, they must evidently be long and narrow. If I

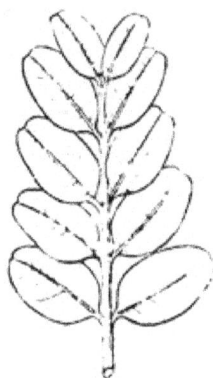

FIG. 65.—Yew.   FIG. 66.—Box.

am asked why they are longer than those of the Yew,
I would suggest that the stem, being thicker, is able
to support more weight. In confirmation of this, we
may take for comparison the Weymouth Pine, in
which the leaves are much longer and the stalk
thicker.

Fig. 66 represents a sprig of Box. It will be
observed that the increase of width in the leaves

corresponds closely with the greater distance between
the points of attachment.

When we pass from the species hitherto considered,
to the Maples (Fig. 69), Sycamores, and Horse
Chestnuts (Figs. 67 and 68), we come to a totally
different type of arrangement. The leaves are placed
at right angles to the axis of the branch instead of
being parallel to it, have long petioles, and palmate

Fig. 67.—Horse Chestnut.

instead of pinnate veins. In this group the mode of
growth is somewhat stiff; the main shoots are per-
pendicular, and the lateral ones nearly at right angles
to them. The buds, also, are comparatively few,
and the internodes, consequently, at greater distances
apart, sometimes as much as a foot, though the two
or three at the end of a branch are often quite short.
The general habit is shown in Figs. 67 and 68. Now,
if we were to imagine six Beech or Elm leaves on these
three internodes, it is obvious that the leaf surface

would be far smaller than it is at present. Again, if
we compare the thickness of an average Sycamore
stem below the sixth leaf, with that of a Beech stem,
it is obvious that there would be a considerable waste
of power. Once more, if the leaves were parallel to
the branch, they would, as the branches are arranged,
be less well disposed with reference to light and air.
A glance at Figs. 67, 68, and 69, however, will show

Fig. 68.—Horse Chestnut.

how beautifully the leaves are adapted to their
changed conditions. The blades of the leaves of the
upper pair form an angle with the leaf-stalks, so as to
assume a horizontal position, or nearly so; the leaf-
stalks of the second pair decussate with those of the
first, and are just so much longer as to bring up that
pair nearly, or quite, to a level with the first; the
third pair decussate with the second and are again

brought up nearly to the same level, and immediately
to the outside of the first pair. In well-grown shoots
there is often a fourth pair on the outside of the
second. If we look at such a cluster of leaves directly
from in front, we shall see that they generally appear
somewhat to overlap ; but it must be remembered
that in temperate regions the sun is never vertical.
Moreover, while alternate leaves are more convenient
in such an arrangement as that of the Beech, where

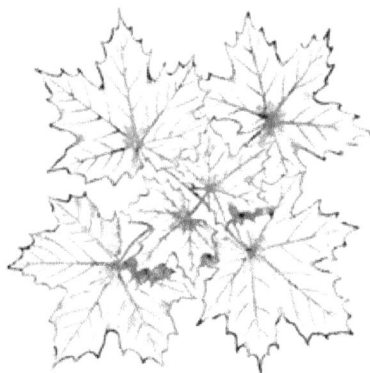

FIG. 69.—Acer.

there would be no room for a second leaf, it is more
suitable in such cases as the Sycamores and Maples
that the leaves should be opposite, because if, other
things remaining the same, the leaves of the Sycamore
were alternate, the sixth leaf would require an incon-
venient length of petiole.

Perhaps it will be said that the Plane-tree, which has
leaves so like a Maple that one species of the latter
genus is named after it (*Acer platanoides*, Fig. 69), has,
nevertheless, alternate leaves. In reality, however, I
think this rather supports my argument, because the

leaves of the Plane, instead of being at right angles to the stem, lie more nearly parallel with it. Moreover, as any one can see, the leaves are not arranged so successfully with reference to exposure as those of the species we have hitherto been considering, perhaps because, living as it does in more southern localities, the economy of sunshine is less important than in more northern regions.

The shoot of the Horse Chestnut is even stouter than that of the Sycamore, and has a diameter below the sixth leaf of no less than $\frac{3}{16}$ of an inch. With this increase of strength is, I think, connected the greater size of the leaves, which attain to as much as eighteen inches in diameter, and this greater size, again, has perhaps led to the dissection of the leaves into five or seven distinct segments, each of which has a form somewhat peculiar in itself, but which fits in admirably with the other leaflets. However this may be, we have in the Horse Chestnut, as in the Sycamore and Maples, a beautiful dome of leaves, each standing free from the rest, and expanding to the fresh air and sunlight a surface of foliage in proportion to the stout, bold stem on which they are borne.

Now, if we place the leaves of one tree on the branches of another, we shall at once see how unsuitable they would be. I do not speak of putting a small leaf such as that of a Beech on a large-leaved tree such as the Horse Chestnut; but if we place, for instance, Beech on Lime, or *vice versâ*, the contrast is sufficiently striking. The Lime leaves would overlap one another, while, on the other hand, the Beech leaves would leave considerable interspaces. Or let us

in the same way transpose those of the Spanish
Chestnut (*Castanea*) and those of *Acer platanoides*,
a species of Maple. I have taken specimens in which
the six terminal leaves of a shoot of the two species
occupy approximately the same area. Figs. 63 and
69 show the leaves in their natural position, those
of the Spanish Chestnut lying along the stalk, while
those of the Maple are ranged round it. In both

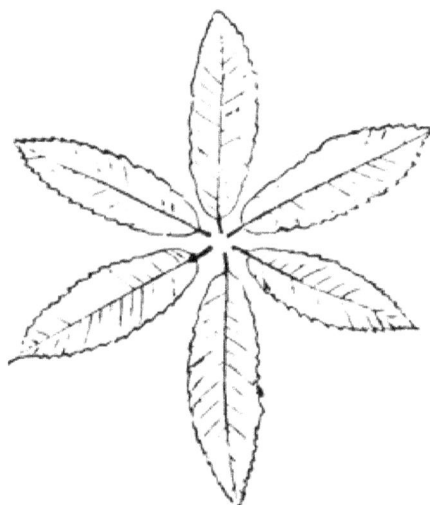

Fig. 70.—Leaves of Castanea.

cases it will be seen that there is practically no over-
lapping and very little waste of space. In the Spanish
Chestnut the stalks are just long enough to give a
certain play to the leaves. In the Maple they are
much longer, bringing the leaves approximately to
the same level, and carrying the lower and outer ones
free from the upper and younger ones.

Now, if we arrange the Spanish Chestnut leaves
round a centre, as in Fig 70, it is at once obvious how

much space is wasted.   On the other hand, if we
place the leaves of the Maple on the stalk of a
Spanish Chestnut at the points from which the leaves
of Chestnut came off, as in Fig. 71, we shall see that
the stalks are useless, and even mischievous as a
cause of weakness and of waste of space; while, on
the other hand, if we omit the stalks, or shorten them
to the same length as those of the Chestnut, as in
Fig. 72, the leaves would greatly overlap one another.

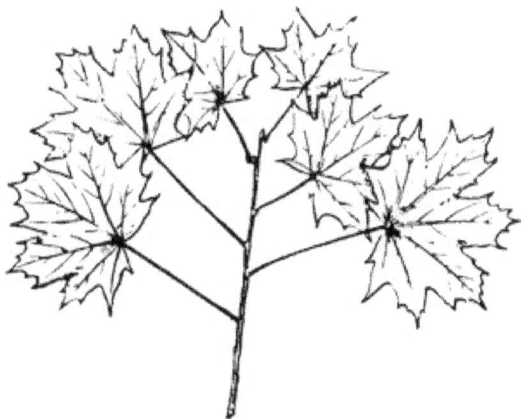

FIG. 71.—Maple leaves on Chestnut.

Once more, for leaves arranged as in the Beech the
gentle swell at the base is admirably suited; but in a
crown of leaves such as those of the Sycamore, space
would be wasted, and it is better that they should expand at once as soon as their stalks have borne them
free from those within. Moreover, the spreading lobes
leave a triangular space (Fig. 69) with the insertion of
the stalk at the apex which seems as if expressly
designed to leave room for the pointed end of the leaf
within.

Hence we see how beautifully the whole form of these leaves is adapted to the mode of growth of the trees themselves and the arrangement of their buds.

Before we proceed to consider the next series of species to which I wish to direct attention, it will be necessary for me to say a few words on the micro-scopical structure of the leaf. Although so thin, the leaf consists of several layers of cells. Speaking roughly, and as a general rule, we may say that on each side is a thin membrane, or epidermis, underneath

Fig. 72.—Maple leaves on Chestnut.

which on the upper side are one or more layers of elongated cells known from their form as " pallisade cells," beneath which is a parenchymatous tissue of more or less loose texture. The leaf is strengthened by ribs of woody tissue. From this general type there are, of course, numerous variations. For instance, many water plants have no epidermis. The structure of the leaf has been described in a number of memoirs by Bonnet, Haberlandt, Areschoug, Stael, Pick, Hein-richer, Vesque, Tschirch, Hentig, and other writers.

If the surface of the leaf be examined with a tolerably high power, small opaque spots will be observed, resembling button-holes, with a thick rim or border composed of two more or less curved cells, the concavities being turned inwards. When dry, they are nearly straight, and lie side by side; but when moistened they swell, become somewhat curved, and gape open. It is difficult to realise the immense number of these orifices, or "stomata" as they are called, which a single bush or tree must possess when we remember that there are sometimes many thousand stomata to a square inch of surface. In a large proportion of herbs the two sides of the leaf are under conditions so nearly similar that the stomata are almost equally numerous on the upper and on the lower side. In trees, however, as a general rule, they are found exclusively on the under side of the leaf, which is the most protected; they are thus less exposed to the direct rays of the sun, or to be thoroughly wetted by rain, so that their action is less liable to sudden and violent changes.

There are, however, some exceptions; for instance. in the black Poplar the stomata are nearly as numerous on one side of the leaves as on the other. Now, why is this? If we compare the leaves of the black and white Poplar, we shall be at once struck by the fact that, though these species are so nearly allied, the leaves are very different. In the white Poplar (*Populus alba*), the upper and under sides are very unlike both in colour and texture, the under side being thickly clothed with cottony hairs. In the black Poplar (*P. nigra*, Fig. 73) the upper and under surfaces

I 2

are, which is not frequent, very similar in colour
and texture. The petioles or leaf-stalks, again, are
unlike ; those of *P. nigra* presenting the peculiarity
of being much flattened at the end towards the leaf.
The effect of the unusual structure of the petiole is
that the leaf, instead of being horizontal as in the *P.
alba* and most trees, hangs vertically, and this again
explains the similarity of the two surfaces, because

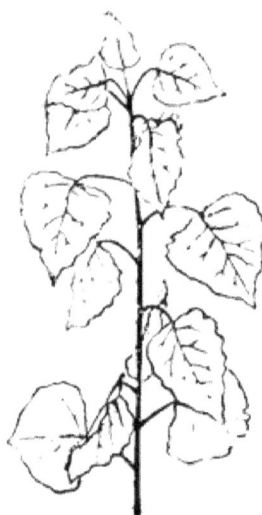

FIG. 72.—Black Poplar (*P. nigra*).

the result is that both surfaces are placed under nearly
similar conditions as regards light and air. Again, it
will be observed that, if we attempt to arrange the
leaves of the black Poplar on one plane, they generally
overlap one another ; the extent is larger than can be
displayed without their interfering with one another.
In foliage arranged like that, for instance, of the Beech,
Elm, Sycamore, or, in fact, of most of our trees, this

would involve a certain amount of waste; but in the black Poplar, as Fig. 73 shows, the leaves when hung in their natural position are quite detached from one another.

Another interesting case of a species with vertical leaves is the Prickly Lettuce (*Lactuca scariola*), while those of other species of Lettuce (*L. muralis* and *L. virosa*) are horizontal. With this position of the leaves is connected another peculiarity, especially well marked in the so-called "compass" plant of the American prairies (*Silphium laciniatum*), a yellow Composite not unlike a small Sunflower, and which is thus named because the leaves turn their edges north and south. This has long been familiar to the hunters of the prairies, but was first mentioned by General Alvord, who called Longfellow's attention to it, and thus inspired the lines in "Evangeline:"

> "Look at this delicate plant, that lifts its head from the meadow,
> See how its leaves are turned no th, as true as the magnet ;
> This is the compass flower, that the finger of God has planted
> Here in the houseless wild to direct the traveller's journey
> Over the sea-like, pathless, limitless waste of the desert."

The advantage of this position, and consequently the probable reason for its adoption, is that in consequence of it the two faces of the leaf are about equally illuminated by the sun ; and in connection with this we find that the structure of the leaf is unusual in two respects. The stomata are about equally abundant on both surfaces, while pallisade cells, which are generally characteristic of the upper surface, are in this species found on the lower one also.

Stahl and Pick have pointed out that even in the same species, leaves growing in shade differ somewhat in these respects from those which are exposed to bright light.

The leaves of the Prickly Lettuce (*Lactuca scariola*) have also, when growing in sunny situations, a tendency to point north and south. Under such circumstances also they have a layer of pallisade cells on each side.

Fig. 74.—*Acacia melanoxylon.*

CHAPTER VI.

HITHERTO I have dealt with plants in which one main consideration appears to be the securing as much light and air as possible. Our English trees may be said as a general rule to be glad of as much sun as they can get. But a glance at any shrubbery is sufficient to show that we cannot explain all leaves in this manner, and in tropical countries some plants at any rate find the sun too much for them. I will presently return to the consideration of other characteristics of tropical vegetation. In illustration, however, of the present point, perhaps the clearest evidence is afforded by some Australian species, especially the Eucalypti and Acacias. Here the adaptations which we meet with are directed, not to the courting, but to the avoidance, of light.

The typical leaves of Acacias are pinnate, with a number of leaflets. On the other hand, many of the Australian Acacias have leaves (or, to speak more correctly, phyllodes) more or less elongated or willow-like. But if we raise them from seed we find, for instance, in *Acacia salicina*, so called from its resemblance to a Willow, that the first leaves are pinnate (Fig. 75), and differ in nothing from those

FIG. 75.—Seedling of *Acacia salicina*.

characteristic of the genus. In the later ones, however, the leaflets are reduced in number, and the leaf-stalk is slightly compressed laterally. The fifth or sixth leaf, perhaps, will have the leaflets reduced to a single pair, and the leaf-stalk still more flattened, while, when the plant is a little older, nothing remains except the flattened petiole. This in shape, as already observed, much resembles a narrow willow-leaf but flattened laterally, so that it carries its edge upwards

and consequently exposes as little surface as possible
to the overpowering sun.   In some species the long
and narrow phyllodes carry this still further by hang-
ing downwards, and in such cases they often assume
a scimitar-like form.   This I would venture to
suggest may be in consequence of one side being
turned outwards, and therefore under more favourable
conditions.

In one very interesting species (*Acacia melanoxylon*, Fig. 74), the plant throughout life produces
both forms, and on the same bough may be seen
phyllodes interspersed among ordinary pinnate
leaves, the respective advantages being, it would
appear, so equally balanced that sometimes the
one, sometimes the other, secures the predominance.

In the case of the Eucalyptus, every one who has
been in the South of Europe must have noticed
that the young trees have a totally different aspect
from that which they acquire when older. The leaves
of the young trees (Fig. 76) are tongue-shaped, and
horizontal.   In older ones, on the contrary (Fig. 77),
they hang more or less vertically, with one edge towards
the tree, and are scimitar-shaped, with the convex
edge outwards, perhaps for the same reason as that
suggested in the case of Acacia.   There are several
other cases in which the same plant bears two kinds
of leaves.   Thus, in some species of Juniper the
leaves are long and pointed, in others rounded and
scale-like. *Juniperus chinensis* has both.

In the common Ivy the leaves on the creeping
or climbing stems are more or less triangular, while
those of the flowering stems are ovate lanceolate ; a

difference, the cause of which has not, I think, yet been satisfactorily explained, but into which I will not now enter.

We have hitherto been considering, for the most part, deciduous trees. It is generally supposed that in autumn the leaves drop off because they die. My impression is that most persons would be very much surprised to hear that this is not altogether

FIG. 76.—*Eucalyptus*—Young.

FIG. 77.—*Eucalyptus*—Old.

the case. In fact, however, the separation is a vital process, and, if a bough is killed, the leaves are not thrown off, but remain attached to it. Indeed, the dead leaves not only remain *in situ*, but they are still firmly attached. Being dead and withered, they give the impression that the least shock would detach them ; on the contrary, however, they will often bear a weight of as much as two pounds without coming off.

In evergreen species the conditions are in many respects different from those affecting deciduous species. When we have an early fall of snow in autumn the trees which still retain their leaves are often very much broken down. Hence, perhaps, the comparative paucity of evergreens in temperate regions, and the tendency of evergreens to have smooth and glossy leaves, such as those of the Holly, Box, and Evergreen Oak. Hairy leaves especially retain the snow, on which more and more accumulates.

Again, evergreen leaves sometimes remain on the tree for several years; for instance, in the Scotch Pine three or four years, the Spruce and Silver Fir six or even seven, the Yew eight, *Abies pinsapo* sixteen or seventeen, *Araucaria* and others even longer. It is true that during the later years they gradually dry and wither; still, being so long-lived, they naturally require special protection. They are, as a general rule, tough, and even leathery. In many species, again, as is the case with our Holly, they are spinose. This serves as a protection from browsing animals; and in this way we can, I think, explain the curious fact that, while young Hollies have spiny leaves, those of older trees, which are out of the reach of browsing animals, tend to become quite unarmed.

In confirmation of this I may also adduce the fact that while in the Evergreen Oak the leaves on well-grown trees are entire and smooth-edged like those of the Laurel, specimens which are cropped and kept low form scrubby bushes with hard prickly leaves.[1]

[1] Bunbury, *Botanical Fragments*, p. 320.

Mr. Grindon, in his *Echoes on Plant and Flower Life* (p. 30), says that "the occurrence of prickles only here and there among plants shows them to be unconnected with any general and ruling requirement of vegetation. We can only fall back upon the principle laid down at the outset, that they are illustrations of the unity of design in Nature, leading us away from the earth to Him who is 'the end of problems and the font of certainties.'" Surely, however, it is obvious that the existence of spines and prickles serves as a protection.

Another point of much importance in the economy of leaves is the presence or absence of hairs. I have already observed that most evergreens are glossy and smooth, and have suggested that this may be an advantage, as tending to prevent the adherence of snow, which might otherwise accumulate and break them down.

The hairs which occur on so many leaves are of several different types. Thus, leaves are called silky when clothed with long, even, shining hairs (Silver Weed); pubescent or downy, when they are clothed with soft, short hairs (Strawberry); pilose, when the hairs are long and scattered (Herb-robert); villous, when the hairs are rather long, soft, white, and close (Forget-me-not); hirsute, when the hairs are long and numerous (Rose-campion); hispid, when they are erect and stiff (Borage); setose, when they are long, spreading, and bristly (Poppy); tomentose, when they are rather short, soft, and matted; woolly, when long, appressed, curly, but not matted (Corn-centaury); velvety, when the pubescence is short and soft to the

touch (Foxglove); cobwebby, when the hairs are
long, very fine, and interlaced like a cobweb (Thistle,
cobwebby House-leek). The arrangement of the
hairs is also interesting. In some plants there is a
double row of hairs along the stem. In the Chick-
weed only one. This, perhaps, serves to collect rain
and dew, and it is significant that the row of hairs
is always opposite to the flower-stalk, which also has
a single row. Now, the flower-stalk is for a considerable part of its life turned downwards, with the
row of hairs outwards. This, perhaps, may account
for the absence of hairs on that side of the stem.

Many leaves are clothed with woolly hairs while in
the bud, which afterwards disappear. Thus, in the
Rhododendron, Horse Chestnut, and other species, the
young leaves are protected by a thick felt, which
when they expand, becomes detached and drops off.
Many leaves are smooth on the upper side, while
underneath they are clothed with a cottony, often
whitish, felt. This probably serves as a protection
for the stomata. In some cases the hairs probably
tend to preserve the leaves from being eaten. In
others, as Kerner has suggested, they serve to keep off
insects—apparently with the special object of pre-
venting the flowers from being robbed of their honey
by insects which are not adapted to fertilise them.
Fritz Müller, to whom we are indebted for so many
ingenious observations, gives an interesting case.
The caterpillar of *Eunomia eagrus*, when about to
turn into the chrysalis (Fig. 78), breaks off its hairs
and fastens them to the twig which it has selected, so
as to form on each side of itself about half a dozen

stiff fences, to protect it during its helpless period of quiescence.

Vaucher long ago observed, though he gave no reason for the fact, that among the *Malvaceæ* (Mallows) the species which produce honey are hairy, and those which do not are glabrous.

Is we make a list of our English plants, marking out which species have honey and which have hairs, we shall find that we may lay it down as a general rule that honey and hairs go together. The exceptions, indeed, are very numerous, but when we

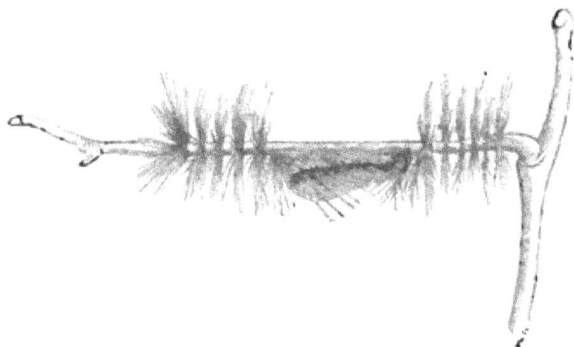

FIG. 73.—Pupa of *Eunomia eagrus.*

come to examine them we shall find that they can generally be accounted for. I have made a rough list of the species in the English flora which have honey and yet are glabrous. It does not profess to be exactly correct, because there are some species with reference to which I was unable to ascertain by personal examination, or by reference to books, whether they produce honey or not. My list, however, comprised 110 species.

Now, in the first place, in sixty of these 110 species the entrance to the honey is so narrow that

even an ant could not force its way in : twenty are
aquatic, and hence more or less protected from the
visits of ants and other creeping insects; thus we
shall frequently find that if, in a generally hairy
genus, one or more species are aquatic, they are also
glabrous—as, for instance, *Viola palustris, Veronica
anagallis, V. beccabunga,* and *Ranunculus aquatilis.*
Polygonum amphibium is peculiarly interesting,
because, as Kerner has pointed out, aquatic specimens are glabrous ; while in those living on land the
base of the leaf produces hairs. Half a dozen are
early spring plants which flower before the ants are
roused from their winter sleep ; about the same
number are minute ground plants to which hairs
could be no protection; three or four are night
flowers ; there still remain a few to be accounted for,
which would have to be considered individually, but
probably the evidence is sufficiently complete to
justify the general inference.

Lastly, I must not omit to mention the hairs which
have a glandular character.

The next point to which I would call attention is
the remarkable manner in which certain forms repeat
themselves. In some cases, there seems much reason
to suppose that one plant derives a substantial advantage from resembling another. For instance,
Chrysanthemum inodorum, the scentless Mayweed,
very closely resembles the Chamomile in leaves,
flowers, and general habit. The latter species, however, has a strong, bitter taste, which probably serves
as a protection to it, and of which also, perhaps, the
scentless Mayweed may share the advantage. These

two species, however, are nearly allied to one another, and I prefer, therefore, to take as an example of mimicry the Stinging-nettle (*Urtica*) and the common Dead-nettle (*Lamium album*). These two species belong to totally different families; the flowers are altogether unlike, but the general habit and the form of the leaves are extremely similar.

How close the similarity is may be seen by the following illustration (Fig. 79), taken from an excellent

Fig. 79.—*Lamium* and *Urtica*.

photograph made for me by Mr. Harman, of Bromley. The plants on the left are true Stinging-nettles; those on the right are the white Dead-nettle, one of which is in flower. So close was the resemblance that, after getting the photograph, I went back to the spot on which they were growing to assure myself that there was no mistake. It cannot be doubted that the true Nettle is protected by its power of sting-ing; and, that being so, it is scarcely less clear that the Dead-nettle must be protected by its likeness to the other. Moreover, though I was fortunate in lighting on so good an illustration as that shown in the figure just when I had the opportunity of photographing it, still every one must have observed that the two species are very commonly found growing together. Assuming that the ancestor of the Deadnettle had leaves possessing a faint resemblance to those of the true Nettle, those in which the likeness was greatest would have the best chance of survival, and consequently of ripening seeds. There would be a tendency, therefore, according to the well-known principles of Darwin, to a closer and closer resemblance. I am disposed to suggest whether these resemblances may not serve as a protection, not only from browsing quadrupeds, but also from leaf-eating insects. On this part of the subject we have as yet, however, I think, no sufficient observations on record.

Ajuga champæpitys, the yellow Bugle, has leaves crowded and divided into three linear lobes, the lateral ones sometimes again divided. They differ, therefore, greatly from those of its allies, and this puzzled me much until one day I found it growing abundantly on the Riviera among *Euphorbia cyparissias*, and I was much struck by the curious likeness. The Euphorbia has the usual acrid juice of the genus, and it struck me that the yellow Ajuga was perhaps protected by its resemblance.

Leaves which float on the surface of still water tend to be orbicular. The water-lilies are a well known illustration. I may also mention *Limnanthenum nymphæoides*, which, indeed, is often taken for a water-lily, though it really belongs to the family

K

of Gentians, and Alisma natans, a species allied to
the Plantains.   In running water, on the contrary,
leaves tend to become more or less elongated.

Subaqueous leaves of fresh-water plants have a
great tendency either to become long and grass-like
or to be divided into more or less hair-like filaments.
I might mention, for instance, *Myriophyllum* ; *Hip-
puris,* or Mares-tail, a genus which among English
plants comes next to Circæa, the enchanter's nightshade, with totally different leaves; *Ranunculus
aquatilis,* a close ally of the Buttercup ; and many
others.

FIG. 80.—*Ranunculus aquatilis.*

Some, again, which, when mature, have rounded,
floating leaves, have long, narrow ones when young.
Thus in Victoria regia the first leaves are filiform, then
come one or more which are sagittate, and lastly
follow the great orbicular leaves.

Another interesting case is that in which the same
species has two forms of leaf (Fig. 80)—namely, more
or less rounded ones on the surface, and a second
series which are subaqueous and composed of more
or less linear or finely-divided segments.

Mr. Grant Allen has suggested that this tendency

to subdivision in subaqueous leaves is due to the absence or paucity of carbonic acid. I have ventured to suggest a different explanation. Of course it is important to expose as large a surface as may be to the action of the water. We know that the gills of fish consist of a number of thin plates, which while in water float apart, but have not sufficient consistence to support even their own weight, much less any external force, and consequently collapse in air. The same thing happens with thin, finely-cut leaves. In still water they afford the greatest possible extent of surface with the least expenditure of effort in the formation of skeleton. This is, I believe, the explanation of the prevalence of this form in subaqueous leaves.

Again, in still air the conditions, except so far as they are modified by the weight, would approximate to those of water; but the more the plant is exposed to wind the more would it require strengthening. Hence, perhaps, the fact that herbs so much oftener have finely-cut leaves than is the case with trees. In the Umbellifers, for instance, almost all the species have the leaves much divided—more, I need hardly say, than is the case with trees. Shrubs and trees are characterised by more or less entire leaves, such as those of the Laurel, Beech, Hornbeam, Lime, or by similarly shaped leaflets as in the Ash, Horse Chestnut, Walnut.

There are, however, many groups of plants, which, while habitually herbaceous, contain some shrubby species, or *vice versâ*. Let us take some groups of this description in which the herbaceous species have their leaves much cut up, and see what is the character of the foliage in the shrubby species.

The vast majority of Umbellifers, as I have just observed, are herbaceous, and with leaves much divided, the common Carrot being a typical example. One European species, however, *Bupleurum fructicosum*, is a shrub attaining a height of more than six feet, and has the leaves (Fig. 81) coriaceous, and oblong lanceolate.

FIG. 81.—*Bupleurum fructicosum.*

The common Groundsel (Fig. 82), again, is a low herb with much-cut leaves. Some species of Senecio, however, are shrubby, and their leaves assume a totally different character, *Senecio laurifolius* and *S. populifolius* having, as their specific names denote, leaves respectively resembling the Laurel and Poplar. In the genus *Oxalis*, again, to which the Shamrock belongs, there is a shrubby species, *O. laureola*, with leaves like those of a Laurel.

I would venture, then, to suggest these consideration

as throwing light on the reason why herbaceous plants so often have their leaves much cut up.[1]

Next let me say a few words on the reasons why some plants have broad and some narrow leaves. Both are often found within the limits of a single genus. I have ventured to indicate the distance between the buds as a possible reason in certain cases. It would not, however, apply to herbaceous

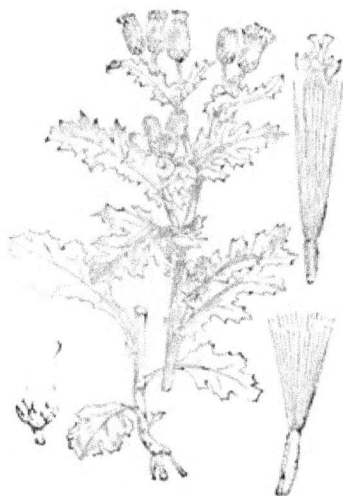

FIG. 82.—*Senecio vulgaris.*

genera such as Plantago or Drosera. Now, *Drosera rotundifolia* (Fig. 83) has the leaves nearly orbicular, while in *D. anglica* (Fig. 84), they are long and narrow, *Plantago media* (Fig. 85) has ovate leaves, while in *P. lanceolata* (Fig. 86) they are lanceolate, and in

---

[1] Mr. Grant Allen, who had been also struck by the fact that herbaceous plants so often have their leaves much cut up, has suggested a different explanation, and thinks it is due to "the fierce competition that goes on for the carbon of the air between the small matted undergrowth of every thicket and hedgerow."

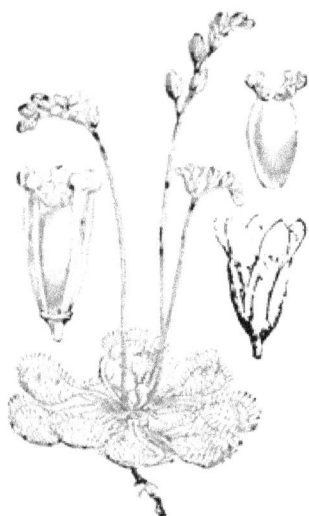

FIG. 83.—*Drosera rotundifolia.*

FIG. 84.—*Drosera anglica.*

FIG. 85.—*Plantago media.*

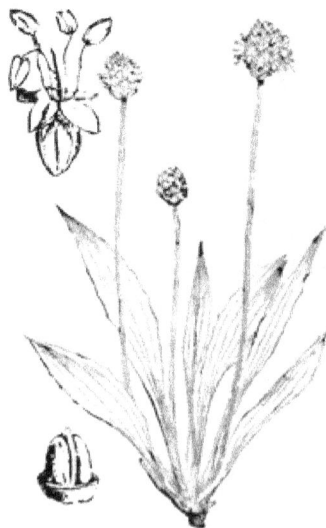

FIG. 86.—*Plantago lanceolata.*

P. maritima nearly linear.　More or less similar cases occur in Ranunculus.

These differences depend, I believe, on the attitude of the leaf, for it will be found that the broad-leaved ones are horizontal, forming a rosette more or less like that of a daisy, while the species with narrower leaves carry them more or less erect.　In the Daisy the rosette lies on the ground, but in other cases, as in Daphne (Fig. 87), it is at the end of a branch.

FIG. 87.—*Daphne*

In hot, dry countries the general character of the vegetation differs from that which prevails in a climate like ours.　There is a marked increase of prickly, leathery, waxy, and aromatic species.　The sap also in many cases is mucilaginous or somewhat salt, which probably tends to check evaporation.　The first two characteristics evidently tend to protect the leaves.　As regards the third, Mr. Taylor[1] in his charming book on Flowers, has pointed to the power

[1] Page 311.

which, as Tyndall has shown, the spray of perfume possesses to bar out the passage of heat rays, and has suggested that the emission of essential oils from the leaves of many plants which live in hot climates may serve to protect themselves against the intensely dry heat of the desert sun.

I am rather disposed to think that the aromatic character of the leaves protects them by rendering it less easy for animals to eat them.

In still drier regions, such as the Cape of Good Hope, an unusually large proportion of species are bulbous. These, moreover, do not belong to any single group, but are scattered among a large number of very different families : the bulbous condition cannot, therefore, be explained by inheritance, but must have reference to the surrounding circumstances. Moreover, in a large number of species the leaves tend to become succulent and fleshy. Now in organisms of any given form the surface increases as the square, the mass as the cube, of the dimensions. Hence, a spherical form, which is so common in small animals and plants, and which in them offers a sufficient area of surface in proportion to the mass, becomes quite unsuitable in larger creatures, and we find that both animals and plants have orifices leading from the outside to the interior, and thus giving an additional amount of surface. But in plants which inhabit very dry countries it is necessary that they should be able to absorb moisture when opportunity offers, and store it up for future use. Hence, under such circumstances fleshy stems and leaves are an advantage, because the surface exposed to evaporation is smaller in proportion than it would be in leaves of the ordinary form. This is, I believe, the reason why succulent leaves and stems are beneficial in very dry climates, such as the Canaries, Cape of Good Hope, &c.

The genus *Lathyrus*, the wild pea, contains two abnormal and interesting species in which the foliaceous organs give the plant an appearance very

FIG. 88.—*Lathyrus niger.*

FIG. 89.—*Lathyrus aphaca.*

unlike its congeners. Fig. 88 represents *L. niger* with leaves of the ordinary type. In the yellow pea (*L. aphaca*, Fig. 89), the general aspect is very different, but it will be seen on a closer inspection that the leaves are really absent, or, to speak more correctly, are reduced to tendrils, while the stipules, on the contrary, are, in compensation, considerably enlarged. They must not, therefore, be compared with the leaves, but with the stipules of other species, and from this point of view they are of a more normal character, the principal difference, indeed, being in size. It is interesting that the young plant has one or two leaves composed of a pair of leaflets, not unlike those of *L. niger*

FIG. 90.—*Lathyrus nissolia.*

The grass pea (*L. nissolia,* Fig. 90) is also a small species. It lives in meadows and the grassy borders of fields, and has lost altogether, not only the leaves, but also the tendrils. Instead, however, of enlarged stipules, the functions of the leaves are assumed by the leaf-stalks, which are elongated, flattened, linear, ending in a fine point, and, in fact, so like the leaves of the grasses among which the plant lives, that it is almost impossible to distinguish it except when in

flower. For a weak plant growing among close grass,
a long linear leaf is, perhaps, physically an advantage ;
but one may venture to suggest that the leaves would
be more likely to be picked out and eaten if they
were more easily distinguishable, and that from this
point of view also the similarity of the plant to the
grass among which it grows may also be an advantage.

In looking at foliage I have often been much
puzzled as to why the leaves of some species are
tongue-shaped, while others are lobed. Take, for
instance, the black Bryony (*Tamus communis*) and
the common Bryony (*Bryonia dioica*). Again, why
are the veins in some leaves pinnate, like those of the
Beech and Elm, and others palmate, as in the Maple
and Sycamore ?

My first idea was that this might have reference to
the arrangement of the woody fibres in the leaf-stalk.
If we make a section of the stalk of a leaf, we shall
find that in some cases the woody fibres are collected
in the middle, while in others there are several distinct
bundles, separated by cellular parenchyma.   My first
idea was that each of the primary ribs of a leaf might
represent a separate woody fibre in the leaf-stalk, so
that leaves with a single bundle of woody fibres
would be pinnate ; those with several distinct bundles
palmate.

The first species which I examined favoured this
view. The Melon, Geranium, Mallow, Cyclamen, and
other species with palmate leaves had, sure enough,
several woody fibres; while, on the contrary, the
Laurel, Rhododendron, Privet, Beech, Box, Castanea,
Arbutus, Phillyrea, and other leaves with pinnate

veins, had one central bundle. But I soon came across numerous exceptions, and had to give up the idea. I then considered whether the difference could be accounted for by the mode of growth of the leaf, and I am still disposed to think that it has some bearing on the subject, though this requires further study.

The next suggestion which occurred to me was that it might be connected with the " prefoliation " or arrangement of the leaves in the bud. The first palmate leaves which I examined were what is called " plicate," or folded up more or less like a fan ; while the leaves with pinnate veins were generally "con-duplicate," or had the one half applied to the other. But, though this was true in many cases, it was not a general rule, and I was obliged to give up this idea also. It then occurred to me to take climbing plants, and see whether I could find any relation between palmate and tongue-shaped leaves on the one hand, and the mode of growth on the other—whether, for instance, the one turned generally up, the other down ; whether the one were generally twining and the other clasping, or *vice versâ*. All these suggestions one by one broke down.

Among Monocotyledons, however, the tongue-shaped preponderates greatly over the palmate form of leaf. With very few exceptions, the forms of the leaves of climbing Monocotyledons are in fact just such as would be obtained by widening more or less the linear, grass-like leaf which is so prevalent in the class.

This, then, raises the question whether the heart

shaped leaf is the older form from which the palmate
type has been gradually evolved.	Let us see whether
we can find any evidence bearing on this question in
what may be called the embryology of plants.	The
Furze, with its spiny prickles, belongs to a group of
plants which, as a general rule, have trifoliate or
pinnate leaves.	Now, if we examine a seedling Furze
(Fig. 91), we shall find that the cotyledons are

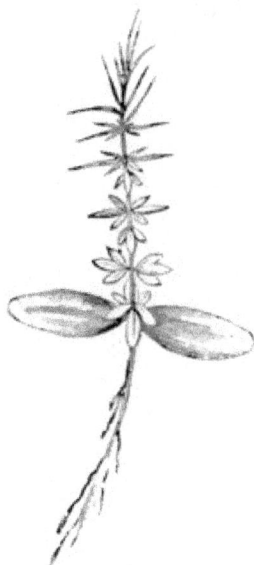

FIG. 91.—FURZE (*Ulex*). Seedling.

succeeded by several trifoliate leaves, with ovate
leaflets.	These gradually become narrower, more
pointed, and stiffer, thus passing into spines.	Hence
we can hardly doubt that the present Furze is
descended from ancestors with trifoliate leaves.	I
have already referred to other cases in which the
young plants throw light on the previous condition
of the species (*ante*, p. 120).

Now we shall have no difficulty in finding cases where, while in mature plants the leaves are more or less lobed and palmate, the first leaves succeeding the cotyledons are entire. This would seem to point to the fact that when in any genus we find heart-shaped and lobed leaves, the former may represent the earlier or ancestral condition.

In support of this it will be sufficient here to give three instances belonging to three distinct families—namely, *Cephalandra palmata* (Fig. 92), one of the

FIG. 92—*Cephalandra palmata.* Seedling.

Cucurbitaceæ ; *Hibiscus pedunculatus* (Fig. 93), belonging to the Malvaceæ ; and, one of the most striking of all, *Passiflora cœrulea* (Fig. 94). Other species of Passiflora show the same passage from entire to trifid leaves, while on the contrary in *P. van Volscenii* even the leaves just following the cotyledons are more or less divided. The advantage of the palmate form may perhaps consist in its bringing the centre of gravity nearer to the point of support.

Broad leaves, moreover, are of two types : cordate,

with veins following the curvature of the edge ; and palmate, or lobed leaves with veins running straight to the edge. The veins contain vascular bundles which conduct the nourishment sucked up by the roots, and it is clearly better that they should hold a straight course, rather than wind round in a curve.

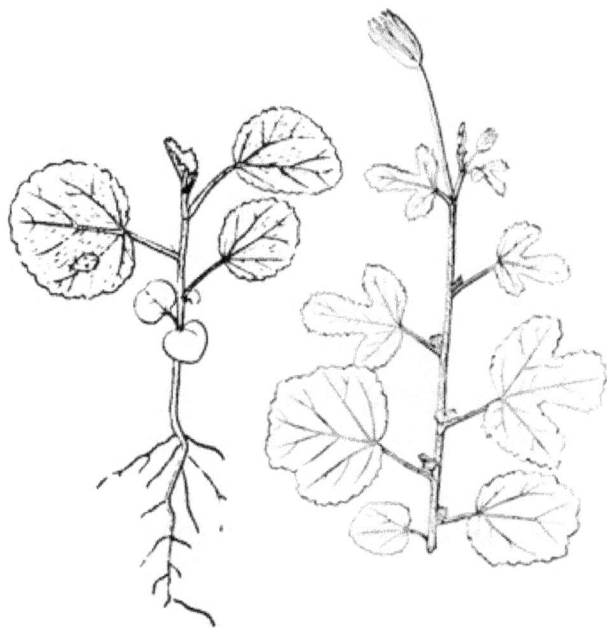

FIG. 93.—*Hibiscus pedunculatus.* Seedling.

Moreover, as the nourishing fluids pass more rapidly along these vascular bundles, the leaf naturally grows there more rapidly, and thus assumes the lobed form, with a vein running to the point of each lobe.

On the whole, we see, I think, that many at any rate of the forms presented by leaves have reference to the conditions and requirements of the plant. If

there was some definite form told off for each species,
then, surely, a similar rule ought to hold good for
each genus. The species of a genus might well differ

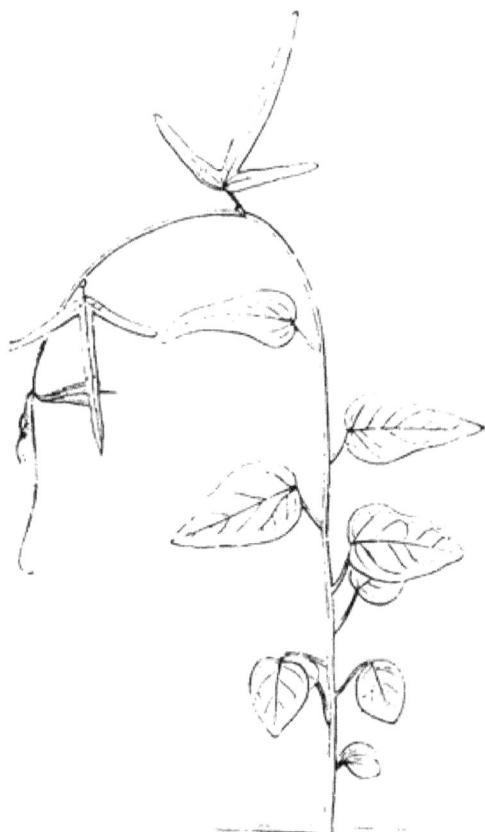

FIG. 91.—*Passiflora cærulea.* Seedling.

more from one another than the varieties of any
particular species; the generic type might be, so to
say, less closely limited; but still there ought to be
some type characteristic of the genus. Let us see
whether this is so. No doubt there are many genera
in which the leaves are more or less uniform, but in

them the general habit is also, as a rule, more or less
similar. Is this the case in genera where the various
species differ greatly in habit? I have already
incidentally given cases which show that this is not
so, but let us take some group—for instance, the
genus Senecio, to which the common Groundsel
(Fig. 82) belongs, as a type well known to all of us
—and look at it a little more closely.

The leaves of the common Groundsel I need not
describe, because they are familiar to us all. This
type occurs in various other species of more or less
similar habit. On the other hand, the fen Senecio
(*S. paludosus*) and the marsh Senecio (*S. palustris*),
which live in marshy and wet places, have long, narrow,
sword-shaped leaves, like those of so many other
plants which are found in such localities. The field
Senecio (*S. campestris*, Fig. 95), which lives in meadows and pastures, has a small terminal head of
flowers springing from a rosette of leaves much like
those of a common Daisy (*Bellis perennis*); a
Madagascar species, as yet I believe unnamed, is
even more like a Daisy. *Senecio junceus* looks much
like a Rush; *S. hypochærideus* of South Africa
strikingly resembles a Hypochæris, as its name denotes.
A considerable number of species attain to a larger
size and become woody, so as to form regular bushes :
S. buxifolius has very much the general look of a
Box, *S. vagans* of a Privet, *S. laurifolius* of a Laurel,
ericæfolius of a Heath, *pinifolius* of a Fir, or rather,
a Yew.

Again, some species are climbers ; *S. scandens*
and *S. macroglossus* have leaves like a Bryony ;

S. araneosus and *S. tamoides* like a smilax or tamus
(Yam); *S. tropæolifolius* like a tropæolum.

Among the species inhabiting hot, dry regions are
some with swollen fleshy leaves, such as *S. haworthii*
from the Cape of Good Hope, and *S. pteroneura*,
from Mogador. *Senecio rosmarinifolius*, of the Cape,
is curiously like a Rosemary or Lavender. Lastly,
some species may almost be called small trees, such

FIG. 95.—*Senecio campestris.*

as *S. populifolius*, with leaves like a Poplar; and
S. amygdaloides, like an Almond.

I might mention, if space permitted, many other
species which, as their names denote, closely resemble
forms belonging to other groups—such, for instance,
as Senecio lobelioides, erysimoides, bupleurioides,
verbascifolius, juniperinus, ilicifolius, acanthifolius,
linifolius, platanifolius, graminifolius, verbenefolius,

rosmarinifolius, coronopifolius, chenopodifolius, lavan-
deriæfolius, salicifolius, mesembryanthemoides, digit-
alifolius, abietinus, arbutifolius, malvæfolius, erodiifolius,
halimifolius, hakeæfolius, resedæfolius, hederæfolius,
acerifolius, plantigineus, castaniæfolius, spiræifolius,
bryoniæfolius, primulifolius, and many more. These
names, however, indicate similarities to over thirty
other perfectly distinct families.

It seems clear, then, that these differences have
reference not to any inherent tendency, but to the
structure and organisation, the habits and requirements, of the plant. Of course it may be that the
present form has reference not to existing, but to
ancient, conditions, which renders the problem all the
more difficult. Nor do I at all intend to maintain
that every form of leaf is, or ever has been, necessarily
that best adapted to the circumstances, but only that
they are constantly tending to become so, just as
water always tends to find its own level.

But, however this may be, if my main argument is
correct, it opens out a very wide and interesting field
of study, for every one of the almost infinite forms of
leaves must have some cause and explanation.

THE END.

RICHARD CLAY AND SONS,
LONDON AND BUNGAY.

BY THE SAME AUTHOR.

PRE-HISTORIC TIMES. As Illustrated by Ancient Remains and the Manners and Customs of Modern Savages. Fourth Edition. 1878. 8vo. 18s. (Williams & Norgate.)

THE ORIGIN OF CIVILISATION AND THE PRIMITIVE CONDITION OF MAN. Fourth Edition. 1882. 8vo. 18s. (Longmans, Green & Co.)

MONOGRAPH OF THE COLLEMBOLA AND THYSANURA. 1871. (Ray Society.)

ON THE ORIGIN AND METAMORPHOSES OF INSECTS. With Illustrations. Third Edition. 1874. Crown 8vo. 3s. 6d. (Macmillan & Co.)

FLOWERS AND INSECTS. With Illustrations. Fourth Edition. 1876. Crown 8vo. 4s. 6d. (Macmillan & Co.)

ADDRESSES, POLITICAL AND EDUCATIONAL. 1879. 8vo. 8s. 6d. (Macmillan & Co.)

SCIENTIFIC LECTURES. 1879. 8vo. 8s. 6d. (Macmillan & Co.)

FIFTY YEARS OF SCIENCE. Being the Address delivered at York to the British Association, August, 1881. 1882. Third Edition. 8vo. 2s. 6d. (Macmillan & Co.)

CHAPTERS IN POPULAR NATURAL HISTORY. 1882. 12mo. 1s. 6d. (National Society.)

ANTS, BEES, AND WASPS. With Illustrations. Seventh Edition. 1883. Crown 8vo. 5s. (Kegan Paul, Trench, & Co.)

ON REPRESENTATION. Crown 8vo. 1s. (Swan Sonnenschein & Co.)

THE PLEASURES OF LIFE. Sixth Edition. Globe 8vo. 1887. 3s. 6d. (Macmillan & Co.)

NATURE SERIES.

ON THE ORIGIN AND METAMORPHOSES OF INSECTS. By Sir JOHN LUBBOCK, Bart., F.R.S , M.P., D C.L., LL.D. With numerous Illustrations. Third Edition. Crown 8vo. 3s. 6d.

ON BRITISH WILD FLOWERS CONSIDERED IN RELATION TO INSECTS. By Sir JOHN LUBBOCK, Bart., F.R.S., M.P., D.C.L., LL.D. With Illustrations. Fourth Edition. Crown 8vo. 4s. 6d.

THE TRANSIT OF VENUS. By G. FORBES, M.A., Professor of Natural Philosophy in the Andersonian University, Glasgow. Illustrated. Crown 8vo. 3s. 6d.

THE COMMON FROG. By ST. GEORGE MIVART, F.R.S., Lecturer in Comparative Anatomy at St. Mary's Hospital. With numerous Illustrations. Crown 8vo. 3s. 6d.

POLARISATION OF LIGHT. By W. SPOTTISWOODE, F.R.S. With Illustrations. Fourth Edition. Crown 8vo. 3s. 6d.

THE SCIENCE OF WEIGHING AND MEASURING, AND THE STANDARDS OF MEASURE AND WEIGHT. By H. W. CHISHOLM, Warden of the Standards. With numerous Illustrations. Crown 8vo. 4s. 6d.

HOW TO DRAW A STRAIGHT LINE: a Lecture on Linkages. By A. B. KEMPE. With Illustrations. Crown 8vo. 1s. 6d.

LIGHT: A Series of Simple, Entertaining, and Inexpensive Experiments in the Phenomena of Light, for the Use of Students of every age. By A. M. MAYER and C. BARNARD. With numerous Illustrations. Crown 8vo. 2s. 6d.

SOUND: A Series of Simple, Entertaining, and Inexpensive Experiments in the Phenomena of Sound, for the Use of Students of every age. By A. M. MAYER, Professor of Physics in the Stevens Institute of Technology &c. With numerous Illustrations. Crown 8vo. 3s. 6d.

SEEING AND THINKING. By Professor W. K. CLIFFORD, F.R.S. With Diagrams. Crown 8vo. 3s. 6d.

DEGENERATION. By Professor E. RAY LANKESTER, F.R.S. With Illustrations. Crown 8vo. 2s. 6d.

FASHION IN DEFORMITY, as Illustrated in the Customs of Barbarous and Civilised Races. By Professor FLOWER. With Illustrations. Crown 8vo. 2s. 6d.

CHARLES DARWIN. Memorial Notices reprinted from "Nature." By THOMAS HENRY HUXLEY, F.R.S.; G. J. ROMANES, F.R.S.; ARCHIBALD GEIKIE, F.R.S.; and W. T. THISELTON DYER, F.R.S. With a Portrait engraved by C. H. JEENS. Crown 8vo. 2s. 6d.

ON THE COLOURS OF FLOWERS. As Illustrated in the British Flora. By GRANT ALLEN. With Illustrations. Crown 8vo. 3s. 6d.

THE CHEMISTRY OF THE SECONDARY BATTERIES OF PLANTÉ AND FAURE. By J. H. GLADSTONE, Ph.D., F.R.S., and ALFRED TRIBE, F.Inst. C.E., Lecturer on Chemistry at Dulwich College. Crown 8vo. 2s. 6d.

A CENTURY OF ELECTRICITY. By T. C. MENDENHALL. Crown 8vo. 4s. 6d.

ON LIGHT. The Burnett Lectures. By GEORGE GABRIEL STOKES, M.A., P.R.S., &c., Fellow of Pembroke College, and Lucasian Professor of Mathematics in the University of Cambridge. Three Courses: I. On the Nature of Light ; II. On Light as a Means of Investigation; III. On the Beneficial Effects of Light. Crown 8vo. 7s. 6d.

Others to follow.

MACMILLAN AND CO., LONDON.